Jules Marcou

The geological map of the United States and the United States

Geological Survey

Jules Marcou

The geological map of the United States and the United States Geological Survey

ISBN/EAN: 9783742820815

Manufactured in Europe, USA, Canada, Australia, Japa

Cover: Foto ©ninafisch / pixelio.de

Manufactured and distributed by brebook publishing software
(www.brebook.com)

Jules Marcou

The geological map of the United States and the United States

Geological Survey

THE GEOLOGICAL MAP OF THE UNITED STATES AND THE UNITED STATES GEOLOGICAL SURVEY.

BY JULES MARCOU.

PREFACE.—This paper is published in order to call attention to the incompetence and extravagance of an administration, bent on entailing difficulties on American Geology which may be perpetuated for centuries in the future.

As Geologists we are all concerned in the doings of the United States Geological Survey, and on us fall the duty to expose the wrong already done, and to prevent, if possible, any further great blunders.

America presents such a splendid field for the progress of Geology, that it is almost unendurable to see it spoiled by unskilful hands, not trained and educated for the work. It is an uncongenial task, and I have hesitated before undertaking it. But after long consideration, I wrote it, with the full knowledge of what is in store for me. I shall be either misunderstood, or disapproved; for it is always unsafe to be too far in advance of your time.

In order to see the effect produced, I asked a friend to read my manuscript, and here is his opinion: "The paper is a strong arraignment of the Survey, and if printed will attract general attention and will be largely read. Of course the Survey men will scout it, and cast all manner of fun upon it, abusing it and its author. But it puts some of the failings of the management in bold light before the scientific world. The strong point, is a call for a larger personal direction committee or commission, etc."

Yes, such will be my reward, but I shall have the conscience of having done my duty, and my whole duty, to American geology. After almost fifty years of geological researches in both hemispheres, my time of usefulness is past, and probably the last service I can render, is to show the actual condition of the Geological Survey and what it ought to be, in order to construct and publish the Geological Map of the United States.

Cambridge, Mass., March, 1892.

INTRODUCTION.—Twelve years have passed since the United States Geological Survey was created by a law approved March 3, 1879, and we have in hand a great variety of publications issued by that Survey, during a full decade. It is time to review the whole question; and see how the work is done, its cost, its progress, its errors, its want, the personnel employed, the plan if any exist, the organization, the methods, and finally to expose the results already realized, and what can be expected in the future of that branch of administration of the Interior Department.

All civilized nations, with hardly two or three half-civilized Mahometan countries excepted, have instituted Geological Surveys. Even Japan, the last accession to modern civilization, has a Geological Survey. The United States was among the first to establish Geological Surveys. States like North Carolina, Massachusetts, New York, Pennsylvania, New Jersey, Virginia, Maine, New Hampshire, Rhode Island, Vermont, Tennessee and Ohio organized Geological Surveys of their own; and the United States government recognizing the absolute necessity to know exactly the mineral value of its Territories voted laws, with appropriations, to make preliminary geological surveys of the Upper Mississippi region and the southern shores of Lake Superior. At the same time all explorations and surveys of the Great West, made by the United States expeditions, sent by the Federal Government, had among their members practical geologists, or at least naturalists of some sort, with the special mission to collect facts and specimens of geology.

The great Civil War from 1861 to 1865, stopped all the researches made under the Federal government; but as soon as it was over, new explorations were organized on greater scale than before; and from 1867 (Dr. F. V. Hayden and Mr. Clarence King), 1868 (Mr. J. W. Powell), and 1869 (Lieutenant Geo. M. Wheeler), to 1879, when the office of Director of the Geological Survey was created, numerous expeditions were organized by the Department of War, the Department of the Interior, and the Smithsonian Institution, which surveyed, more or less carefully, about one third of the region west of the Mississippi and Missouri rivers. Being in a very rough country, hardly settled except on a few lines of roads, along a single railroad track (the Union and Central Pacific), some trading posts and Military forts, all those surveys are only preliminary works; and although valuable, their reports are necessarily very incomplete and rather first sketches, than anything else.

However, the great difference between those surveys of the period comprised between 1867 and 1879, and those prior to the great Civil War, is that instead of being undertaken only for the special *desideratum* of knowing more about the public domain, and simply of being purely scientific explorations, the former became from the first, more or less, scientific-political organizations. That they drifted from purely scientific motives into the political arena, was due mainly to each one of the Directors of those explorations, all four, more or less. scientist-politicians.

Until then all the scientific work undertaken by the government of the United States, was done entirely by men of scientific capability and knowledge. Such is the work and organization of the Coast Survey, the Smithsonian Institution and the Engineer Corps of the Army. And it can be said with justice, that all those organizations have given excellent results, in every desirable way. No scientist-politicians have ever found the roads to absorb and direct any of them. Scientific they have been created, and purely scientific they are to this day. Of course, a few rivalries to obtain first places and best positions in those organizations have always existed; but in general the best men have been chosen for Directors or Chiefs.

Unhappily it is not so with the United States Geological Survey. From the beginning a series of intrigues, rarely equalled among politicians in Washington, were started, always with an appearance of endorsment by a certain number of scientists, more or less interested in the final settlement.

As soon as the government of the United States came to the decision of having a Geological Map constructed for its use, the Secretary of the Interior ought to have tried at once to get the best plan and information on the subject. Instead of listening to the interested solicitations of Messrs. Hayden, King and Powell, the Secretary ought to have called for the opinions of geologists, having much more experience; and even to have obtained from foreign governments their organizations of Geological Surveys and how they have proceeded. Besides, we have in this country a good example to follow, with the creation of the Smithsonian Institution; for which plans were suggested and offered to the government by competent persons. A sort of concourse. In the case of the Smithsonian, the best plan proposed was chosen, and notwithstanding its author — Professor Joseph Henry — was not a candidate to direct the whole concern, he was appointed nevertheless Secretary of the new institution.

The following plan to construct a Geological Map of the United States and organize a Geological Survey, is submitted to those able to judge, or at least to those who are accustomed to handle scientific subjects.

A DIRECTOR GENERAL OF THE U. S. GEOLOGICAL SURVEY. —The election of the first Director is of primary importance, because on the choice of this single officer will depend the future good name, and success, and usefulness of the Geological Map of the United States. A board composed of competent men, able to appreciate geological works, as well as administrative-scientific organizations ought to be appointed, and its most important duty would be to find a proper person to direct the new Bureau. The choice must fall not only on an honest man, but above all, on an able geologist well acquainted by previous work with the geology of North America as it may stand at the moment of the election, and having also a sufficient practical knowledge of the standard geology of Europe to be able to direct all researches : stratigraphic, paleontologic, lithologic, coloring of maps and sections, according to the best methods ; and to arrive at right conclusions on every important subject of classification, nomenclature and general synchronisms and equivalencies in both hemispheres. If such a man should not exist, it should be the duty of the Board to choose two geologists, most promising by previous work, made on the geology of some part of the United States, and send them to Europe at the government's expense, to study there practically the geology of England, France and Germany, during one year at least. And after their return, see which is the best of the two and appoint him Director and the second Assistant-Director.

For such a responsible position before the scientific world geologist-politicians ought to be avoided by all means, as well as lobbyists or hangers on for situations on the government pay. Washington, even before the great Civil War, but a great deal more since, has been always a hot bed of persons eager to get a comfortable living from the United States Treasury, and ready to try to get any office, old or new, no matter as regard to their qualification, which according to their standard is only a secondary object.

THE GEOLOGICAL MAP OF THE UNITED STATES. — The first want and principal object is to construct a geological map, on a scale not too great — in order to be easily consulted and handled — to be used by government in all general questions which may arise for Industry, Commerce, Agriculture, Forestry, Mining, Railroad, etc. That map should be correct, in regard to chronology, determination of rocks represented in each geological basin, and also in regard to exact limits of formations. For without these essentials a geological map is less than worthless, it is misleading.

In such a great country, covering the breadth of a whole continent from the Atlantic to the Pacific shores, a general geological map must necessarily be so limited in dimensions as to show at a glance the information wanted, and allow a comparison between the different area or stretches of country. Practically a map in ten or twelve large sheets, for the whole United States, is sufficient; and it is desirable to get it as promptly as possible ; say in the space of twelve or fifteen years at most. The scale 1 : 2,000-000 will give a map of the right dimension. The United States Engineer Corps possess a map at that scale, for the country west of the Mississippi river, in six large sheets ; in extending it to the Atlantic borders, it will add four other sheets, making in all ten sheets. It may be that a map at the scale of 1 :1,000,000 may be chosen, as more convenient for area of mountains, or for complicated regions where several formations are sometimes crowded together ; however, one at the scale of 1 :2,000,000 colored geologically would supply all the wants for general purposes, either practically or scientifically, and it may be constructed in less time and at less expense. The choice of the scale is only a secondary question, which may be reserved. A reduction of that map to one or two sheets only, would be in general use at once all over the country for schools, colleges, census, mining, agriculture, etc.

THE CONSTRUCTION OF THE MAP. — After all the essays made in different countries, the best method is to divide the work in the following way. First of all, it is impossible on a general map to mark distinctly the superficial deposits, such as the Recent and the

Quaternary formations, which cover more or less, almost all the older rocks; and a special map is required exclusively for those two formations, forming one single period: the human age or Modern period. In fact, such superficial and easy geology is considered justly as belonging to Physical Geography, and may be dispensed with in a Geological Survey. However, considering the immense surfaces covered by old glaciers, the importance of the Mississippi river alluvial, and of the coast formations, it will be perhaps better to make a special division in the United States Geological Survey for Modern rocks and deposits; only that division must be reduced to a few observers, two or three at most. A single volume of explanation will be sufficient for the superficial deposit map.

The second map, and by far the most important, as it is the true Geological Map, may be worked by formations or systems. The Geological Survey Corps should consist then of as many divisions as there are systems adopted for representation on the map. All practical geologists, who have worked conscientiously and with a clear understanding of their duties, have arrived at the following great divisions for the rocks composing the structure of our globe: Tertiary, Cretaceous, Jura, Trias, Dyas, Carboniferous, Devonian, Silurian, Cambrian or Champlain, Taconic, Crystalline rocks, Volcanic and Eruptive rocks.

The Tertiary is divided in two: the upper part comprising what has been called by Lyell, Pliocene and Miocene ; and the lower part comprising the Oligocene, Eocene and the Calcaire pisolitique of Paris. The Tertiary Division should be composed of one geologist and three assistant-geologists; all four well trained in inverterbrate paleontology and having a practical knowledge of the Tertiary of France and Switzerland.

The Cretaceous division ; one geologist and two assistants ; all well trained in inverterbrate paleontology and in a practical knowledge of the Cretaceous series of England, France and Switzerland.

The Jura division ; one geologist and one assistant, both trained in inverterbrate paleontology and in the jurassic series of England, France, Würtemberg, and Russia.

The Trias division ; one geologist and one assistant ; both trained in invertebrate paleontology, and one of the two having a good knowledge of paleophytology, and both having studied practically the typical series of England, France, Würtemberg and the Austrian Alps.

The Dyas division; one geologist and one assistant; both trained in invertebrate paleontology and one of the two in paleophytology, and knowing practically the typical series of England, Saxony and Russia.

The Carboniferous division; one geologist and three assistants; two well trained in inverterbrate paleontology and one a good specialist for paleophytology; and all four well acquainted with the typical Carboniferous of England, Ireland, Belgium and Russia.

The Devonian division; one geologist well trained in invertebrate paleontology and well acquainted with the typical devonian of England, Belgium, the Rheinish provinces of Germany and Russia.

The Silurian division; one geologist, well trained in invertebrate paleontology, and knowing practically the typical silurian of Great Britian, Bohemia and Russia.

The Cambrian or Champlain division; one geologist and one assistant; both trained in invertebrate paleontology and acquainted with the typical localities of Wales, Bohemia and Scandinavia.

The Taconic division; one geologist and three assistants; well trained in invertebrate paleontology and with a good practical knowledge of the typical primordial localities of Wales, Bohemia, Scandinavia and Russia.

The Crystalline rocks division; one geologist and one assistant; both well acquainted with lithology and oragraphy.

The Volcanic and eruptive rocks, one geologist well trained in microscopic lithology.

Making twelve geologists and sixteen assistant-geologists. Adding the three of the division of the Recent and Quaternary deposits; and also the Director and Assistant-Director; it is a total of thirty-three officers for the United States Geological Survey.

The administrative part should be performed by a clerk with three assistants; and one Librarian with one assistant. The Librarian and his assistant being good geologists and paleontologists, in order that they may fulfil their positions with intelligence and capacity.

Each division should work steadily, all the year round in the field, pursuing the system of rocks which it has been assigned to, until it has covered the whole field. If a party come upon a field containing rocks belonging to another system, it will be its duty to signalize it at once to the other parties, through the Assistant-Director.

The Director and Assistant-Director, should have the charge, besides the direction of practical work, to concentrate and unite together on the manuscript Geological Map of the United States, the results arrived at by each party.

TIME REQUIRED TO CONSTRUCT THE GEOLOGICAL MAP OF THE UNITED STATES. — Ten or twelve years at most, is a sufficient lapse of time to execute all the researches in the field necessary to fill up the Geological map, with exact data. Then three other years to prepare the map, sections, reports and descriptions of fossils. And in fifteen or sixteen years, at most, we should have a good Geological Map of the United States, with Explanations, Characteristic fossils and descriptions of the rocks.

If any of the geologists, and assistant-geologists — all carefully chosen, among scientists, having each one a scientific record — many have not made previous studies of the typical European localities, where each system of rocks was worked out first, and is considered in practical geology as the standard; they ought to be sent to Europe and study carefully there, in the field — not in Museum or private cabinet — the system of rocks which they may have been assigned to. Six or eight months of hard work in England, France, Switzerland, Germany, Scandinavia and Russia, according to the standard localities required for each system, would suffice to give each one of them a store of knowledge and a base, which cannot be supplied in any other way.

It is the result of practical geology all the world over, that good work, in geology, cannot be expected to be complete and reliable, without comparison, synchronism and systematic equivalency with typical localities, taken as standards; and descriptions of those typical localities, as complete and excellent as they may be in print, are not sufficient and cannot replace the study *in situ* of each one of them. It is as necessary for a geologist to see for himself and study the typical localities, as it is for an artist to go to the great masters and pass months or years there sometimes, in studying their methods, rules and processes, beside having the advantage, beyond estimation, of contemplating in all their perfections the *chef d'œuvres* or typical models, regarded as the most perfect in existence.

The expenses incurred by the government to send out to European typical localities a certain number of geologists, will be fully repaid by the value of the work done on American geology. Instead of creeping in the dark, as it has been, and is still too often the case, notwithstanding the good will, capacities and enthusiasm of some geologists, our official geologists will have a solid base to rest upon, and not be all the time hesitating, or going from one blunder to another, as has been too often the case during the last forty years.

Against the objection to sending American geologists to Europe, for special studies, we have precedent in our government which can be quoted. For instance, officers of the Army have been sent repeatedly to study and make reports on improvements in the material of war, tactics and equipments; also some Navy officers are, now and then, detailed on the same errand.

PUBLICATIONS. — It would be the duty of the Director and Assistant-Director to exert a severe and thorough scientific control on all that relate to the publications of the Survey. The Reports and the Geological Map should be presented in manuscript to the Board of Regents or Council, composed in addition to the usual official members such as: the Secretary of the Interior, the Director of the U. S. National Museum, three U. S. Senators, and three members of the House of Representatives; of ten citizens all well known geologists or paleontologists representing all the diversity of opinion existing on the different questions relating to American geology. Several executive Committees of Publications should be appointed by the Council, composed, each one, of three persons — two of whom should always be competent geologists on the subject treated — to which committees every manuscript should be referred before its acceptance for publication.

The Annual Report should be exclusively administrative. If some very important discoveries which may affect the progress of American geology are made, then the Director may ask the officers of the Survey who may have made the discoveries to write very short résumés of them, which shall be presented to the Annual Meeting of the Council; reported upon by Committee, and if accepted, added as an Appendix, at the end of the Administrative Report. But in no case should a member of the Geological Survey be allowed to read or present a paper, relating in any way to his observations made, when on duty in the Survey, before any scientific societies or academies, or to any scientific periodicals.

The bibliographical works are part of the duties of the Librarian and his assistant; and the officers working in the field should be left entirely to their practical observations, until they shall have finished; working during the summers on the high plateaux in the Mountain ranges, or in the northern band of the United States, and during the winters in the Southern and Middle States, as far as practicable. No office work should be done by the different divisions until all their notes have been taken in the different areas where their systems crop out; and then after ten, eleven or twelve years, according to the time they have finished their practical surveys, they should report to Washington, and begin office work.

The edition of the Publications for the use of the Geological Survey need not be, in any case, above twelve hundred copies. A number amply sufficient for all purposes. The only exception may be, for a reduced map in one or two sheets, of the final Geological Map of the United States, which may be printed to three thousand copies.

EXPENSES. — The expenses for such a survey must be kept in reasonable limit; say, between $150,000 to $200,000 a year. During sixteen years, it would make a total of three millions or $3,200,000; for which we should possess a good Geological Map of the United States; sufficiently reliable for all practical objects, which may be wanted by our government in general and by all the citizens. Of course, there might be occasions which would require detailed surveys of some special localities; then the work should be executed by the United States Government, if of public interest, and under a special law, or by the State if of a local interest only, or by individual or companies if of private interest. In all cases the United States Geological Survey as a corps must be kept free from interferences with geological work done by States and corporations.*

_____

* Other plans may be easily proposed; but they will be less practical, more expensive, and above all not so efficient for arriving promptly at the result wished for;

Now let us make a review of what has been already done by the United States Geological Survey and its organization.

THE ACTUAL UNITED STATES GEOLOGICAL SURVEY.—Without entering into many details, well known by all those interested in American geology, I shall quote only the main facts.

DIRECTOR CLARENCE KING. — The law creating the office of Director of the Geological Survey was enacted and approved, March 3, 1879. In the too short time of eighteen days, the President nominated on March 21, as Director, Mr. Clarence King. It was a politico-scientific nomination; a great deal more political than scientific. Less than two years after, Mr. Clarence King sent his resignation, under the very curious plea, that he "can render more important service as an investigator," and that his work as the head of an executive bureau "leaves him no time for personal geological labors." After that failure, for it cannot be called anything else, the President, acting under interested and very unwise advices, with even more haste than in the first appointment, nominated, only one day after receiving the resignation of Mr. King, the 14th of March, 1881, Mr. John W. Powell, as Director of the Geological Survey. It was simply a transfer arranged beforehand.

When the law was enacted by Congress, in March 1879, there were in Washington three candidates for the position of Director; no one of them was well fitted for the work; being all three merely scientist-politicians. One was already provided with a position in the government employ, and ought to have been satisfied with the direction of the Ethnologic Bureau; for to conduct such scientific researches in Ethnology and Anthropology requires all the talent, knowledge and time that a man of great value as a scientist can bestow on it.

Of those three scientist-politicians, the least objectionable was Dr. F. V. Hayden. Though not a good geologist, by any means, he had at least an enthusiasm in geology. To be sure he was not a competent and exact observer, and was wanting in seriousness and dignity, being too nervous and too eager to make

---

and they will interfere, more or less, with the States Geological Surveys which ought to be avoided by all means.

Of course, revisions with corrections, and new editions of the Geological Map of the United States and of its Explanation will be wanted; and the organization would be retained in part, with a reduction of two-thirds of the officers employed. But it is a question to be discussed when the time comes for it.

a good and friendly impression on everybody he met with. But with all his faults, defects and want of knowledge, he was the best of the three candidates.

If the government had been well advised, the choice would have fallen on some one, not a candidate and entirely outside of the ring of Geologists, more or less connected with all the errors and bad geology implanted on American Geology since 1846.

By some error of judgment and an entire ignorance of geological science, the choice was the most unfortunate imaginable. Not only the government nominated one of the three scientist-politicians candidates, but its first choice fell on one who is hardly a geologist at all. To be sure Mr. Clarence King is a gentleman, a graduate of one of our oldest Colleges and has received a cosmopolite education. He is a pleasant writer of articles for Reviews, and the author of a popular volume : " Mountaineering in the Sierra Nevada," giving an account of life among the gold seekers and herders, as it was fifteen years after the discovery of gold and the great rush to California. But all those accomplishments, do not make him a geologist.

As director of the exploration of the fortieth parallel he has given his exact measure of his capacities and scientific knowledge. Unknown as geologist, all that was known of him was that he had passed several years in California, as an assistant of of the Geological Survey of that state—a survey which had become a byword among surveys badly managed, and a choice example of a failure of a State Geological Survey. Mr. King has taken to geology, not because he was a good geological student and had truly in him the stuff to make an accurate and able geological observer, but simply for private family reasons. During his stay in California, Mr. King made friends and obtained from some of them letters of recommendation for men high in office at Washington. Armed with those letters, more especially some from Colonel R. S. Williamson, chief of the Engineers on the Pacific coast, with whom he agreed, before leaving San Francisco, to share with him the Survey of the line of the proposed Pacific Railroad by the fortieth parallel, already in course of construction, Colonel Williamson having in charge the Topographic and Geographic Survey and Mr. King only in charge of the Geology and the Natural History ; armed with those letters, as I say, Mr. King came to Washington and delivered them into the hands of General A. A. Humphreys, chief of Engineers. Acting at once on those recommendations, General Humphreys arranged a well balanced plan for the execution of the proposed survey, the Topographical work being placed in the hands of the Corps of Engineers, and Mr. King having only the direction of the Natural History (Geology, Mining, Paleontology, Botany and Zoology). The draft of the law was accepted and ready to come next day before Congress, as drafted by General Humphreys, when by some underhand negotiations, kept carefully secret, the draft of the law was changed during the evening in such way that Mr. Clarence King was to direct not only the Natural History part of the exploration, but also the Topographic and Geographic Survey. General Humphreys knew nothing about it, until the matter had become a law. Too gentle and kind to take offence at the strange process, General Humphreys accepted matters as they were, and did nothing to impede King in his work; on the contrary, contenting himself with endorsing all the vouchers of the expenses incurred by that exploration.

From his first appearance in Washington, Mr. King made use of what is called the "third house," or the lobby. With the creation of the office of Director of the United States Geological Survey, he acted in the same manner and with the same success. Very likely, Mr. King belongs to that school who say that the end to attain, justifies and excuses the means by which it is reached.

Scientifically, the Exploration of the fortieth parallel is mainly due to Mr. King's assistants and associates. The first volume claimed as his special share in the publication of that Survey, is simply a rehearsal of the works made by others; he extended with amplification and a certain facility of a writer of Reviews, as he is, the observations made by others, more especially those made by Messrs. S. F. Emmons and A. Hague, and without sufficient reference to their authorship, he appropriated largely work done by others.

As Director of the U. S. Geological Survey, during the first eighteen months, he did no original geological work at all; and as to "the more important service to science as an investigator," as he claimed in his letter of resignation to the President of the United States, he has absolutely published nothing, or made known, in any form or way to the scientific world, a single fact of his original geologic investigations. It is true, he reserved to himself the first Monograph of the U. S. Geological Survey and the apparition of the volume was announced as to be devoted by Mr. King to his observations on "The precious metals," (Gold and Silver) ; but after ten years of expectation, the scheme was finally abandoned, and the first Monograph appeared simply as a description of the physical geography of the Great Salt Lake area, by Mr. G. K. Gilbert, Washington, 1890.

The example given by Mr. King, so totally different from every other scientific appointment made before by the United States Government is profoundly regretable. and is anything but gratifying to American Geology.

DIRECTOR J. W. POWELL.—Although already Director of the Bureau of Ethnology, Mr. Powell was nominated Director of the Geological Survey, in addition to his first appointment. If Mr. King was objectionable on account of his very limited knowledge of Geology and total ignorance of Paleontology, it is even more so with Mr. Powell. Not trained as a Geologist, Mr. Powell, thanks to the great kindness and good heart of Professor Joseph Henry, obtained the means to make explorations in the region of the upper Colorado river, under the direction of the Smithsonian Institution. Two reports on the Colorado river and the Uinta Mountains, published at Washington in 1876, have given the result of his investigations.

The geological parts of those reports, are rather meagre and do not show any fitness in their author to direct the Geological Survey and the making of the Geological Map of the United States. In fact, the only geological map accompanying these reports, entitled "Green River from the Union Pacific Railroad to the mouth of the White river" is simply a reproduction. No name of author is given, except that it is marked as belonging to "the Second division of the U. S. Geological and Geographical Survey of the territories"; J. W. Powell, Geologist in charge. It would seem, then, that the map ought to be credited to Mr. Powell. But if we compare it with the two maps of the Green river basin, by S. F. Emmons, in the "Atlas of the fortieth Parallel," every one will be struck with their perfect similarity and even identity. The original map of Mr. Emmons is above discussion as to authorship, and consequently the Green River geological map belongs to him. It is impossible to have such identity, without a communication of the original map to Mr. Powell.

Since his appointment in 1881, ten years have elapsed, and Mr. Powell has not published either a Geological map, or a Geological memoir during all the decade.

By successive additions and constant increase, the Geological Survey, instead of confining its scope to the construction of the Geological map of the United States, with a clear and sufficient explanation, has embraced all sorts of subjects, absolutely out of the line of Geological researches.

THE TOPOGRAPHICAL SURVEY.—First the Geological Survey has undertaken to construct a Topographical map of the United States. There is no doubt, that a good Topographical map, at a scale of at least 1 :100,000, is much wanted and ought to be constructed and issued by the United States government. But the methods of working, and the whole subject, is entirely different and altogether another department of scientific investigation.

Geodesy and Topography have nothing to do with Geology; and all the "personnel" engaged in surveying, plotting and issuing a topographical map of the United States, is absolutely devoid of geological knowledge. The Union in the same Bureau of a Geodesic and Topographic Survey with the Geological Survey is a great drawback on both Surveys, and most especially on the Geological Survey. Instead of helping one another, on the contrary, it complicates the whole work, without the smallest advantage. It is as difficult scientifically, to make triangulations and represent the topography of such an immense country as ours, as to trace its geological map; and to direct one or the other survey requires scientific acquirements and capacities of the first order, which are never united in one man, even if that man is a man of genius. It is materially and scientifically impossible for a man, however well educated and clever, to direct in a satisfactory way, both surveys.

The result of the amalgamation of the Geological Survey with the Topographical Survey has been, so far, hurtful to both; and the longer they remain confounded together under the same Director, so long their value will be diminished and of a low standard. The appropriations required annually for each one, after starting on a point of equality, will diverge and is diverging

more and more, until the Topographical Survey will cost four
and five times more than the Geological Survey.

So far the Topographical Survey has used three different
scales: 1 :62,500 — 1 :125,000 — 1 :250,000, and the work done
is generally far below those executed in England, France,
Belgium, Switzerland, Italy, Germany, etc. Not one of the
sheets yet published — more than 300 — can be called a good
topographical map, and be given as a standard to compare with
foreign maps. In some cases they have been rejected and the
States have remapped the same areas. To be sure, it is better
than nothing; but with the best desire to accept as a boon the
bestowal by the government of the first Topograhical map it has
undertaken, the fact is, that the sheets issued are only acceptable
as a first and most incomplete draft, of what ought to be " The
Topographical Map of the United States." The scale of
1 :62,500 is most awkward and very objectionable; and it ought
not to have been used and adopted by the Director.

IRRIGATION SURVEY. — But this is enough on the " Branch
of Geography " of the Geological Survey, as it is called; let us
pass to another subject. As if it was not already enough for
Mr. J. W. Powell to direct three great Surveys — Ethnology,
Geology and Topography — a fourth Survey has been added in
1888, under the title of " Irrigation Survey,"* to construct reser-
voirs, ditches and canals; a whole hydraulic system of works,
covering surfaces as great as the area of one-third of Europe.
And now we have added another branch to the Geological Survey, a branch having nothing to do with Geological researches
and absolutely independent, requiring the highest knowledge of
the Hydraulic Engineer.

Mr. Powell is not a hydraulic engineer, nor even an engineer
at all, no more than he is not a Topographical engineer, or a

---

* The concentration of four great Surveys, into the hand of a single Director,
allows great margin to favorise one at the expense of another. For instance the
Bureau of Ethnology, being the pet Survey of Mr. Powell, absorbs a part of the
Geological Survey building and most of the time of the artists and draughtsmen.
Out of the money appropriated in 1891, amounting to $750,000, only a small portion
is devoted truly to pure geological work, and even a large part of that goes in salaries not specially provided for otherwise. In fact the name Geology is used before
Congress, to nurse all sorts of organizations, which otherwise will fail to get sufficient
appropriations for their strong appetite; and the Geological Map of the United
States instead of receiving the lion share of the appropriation drawn in its name is
reduced to a meagre portion: although already too great for the poor result arrived
at, each year.

Geologist, or a Paleontologist, or an Ethnograph. How to explain such ambitious dreams, and such a bold estimate of his own value? For never such a conceited man has existed in science, anywhere. Mr. Powell has not received a regular education, either in any University, or at West Point, or at the Naval school, or in any Technological school. He has never published a single paper which may be called a good work, and which is quoted as such by scientists. It is said that Mr. Powell was formerly a schoolmaster in a village of Illinois. We all know that schoolmasters in small villages are obliged to teach all sorts of matters to their pupils, and consequently they have all some very slight and elementary notions of geography, land survey, mathematics, linguistics, natural history, etc.

Living among farmers and mechanics, and regarded as the learned man of the community, it is easy to conceive that a schoolmaster may become such a conceited man that he may believe himself able to do anything, and even to conduct any scientific or literary work. But the great difficulty lay in getting hold of any great Bureau of the Government in Washington. Unhappily, in our country, a certain class of politicians have set an example to divide spoils, which has turned the heads of many ambitious men desirous of getting a good position at the expense of the United States Treasury. Science at first escaped from the grasp of politicians; but it was too tempting for half-scientists, such as may be expected from the class of schoolmasters, not to follow the ways used in that part of the Capitol called the lobby. After some years of essay Mr. Powell has become such an expert, that he is regarded in Washington as the leader and most influential member of the third House. No wonder that the Geological Survey of the United States can obtain such large appropriations, and that a man can dispose of money without any scientific control of any sort and at his own will, provided that his annual accounts are properly audited at the Interior and the Treasury departments.

It is almost incredible that a man should have the audacity to assume such scientific responsibilities. It is unique in the world. The only explanation and excuse is that Mr. Powell is not a scientist, in the full meaning of the term.

But all those scientifico-political successes do not make a good geological map of the United States, which is the object of the oaws enacted in 1879; and a short review of what has bu doene during the decade may give an idea of how the work is carried in, what has been done, and what remains to be done.

PLAN OF THE UNITED STATES GEOLOGICAL SURVEY.— Strange to say, there is no plan. Everybody,—including "heads of divisions," called also "chiefs of divisions and heads of independent parties,"—does pretty much what he pleases, and when he pleases. We have numerous examples already of transfers of heads of divisions from one division to another; of the creation of quite in formidable number of new divisions; of a head of division retir- elig from the Geological Survey to accept a more lucrative pos- tion with a great company—then, after the failure of the company, returning simply to the United States Geological Survey and being placed at once at the head of a new division. The main object —the construction of the Geological Map of the United States is entirely forgotten; no one ever speaks of it; the Directors like the rest, seems entirely absorbed by other interests, and a complete stranger to what is going on. Only now and then, when the fiternational Congress of Geologists meets, in order to make a figure and a show, we see appear tables of "Colors for Geologic Cartography"; a "General Scheme adopted by the United States Geological Survey for Nomenclature"; "Standards for Geological Cartography" with "Illustrative Patterns"; "Conventional Char- acters for Diagrams"; "Methods of Geologic Cartography in use by the United States Geological Survey"; "Map of the United States exhibiting the present status of knowledge relating to the areal distribution of geologic groups" (preliminary compilation), compiled by W. J. McGee, 1884; and finally a "Conference on Map Publication," with conventional symbols for geologic maps and geologic sections, held at Washington on January 28 to 31, 1889. "The Illustrative Plates" and the report of that "Confer- ence" were distributed at the meeting of the International Congress of Geologists, held at Washington, in August, 1891, as the last work of the Geological Survey of the United States on geological maps.

Let us consider the last two publications quoted above; for we have there the occasion to get a glimpse at what may be, if not a plan, at least a tendency of the United States Geological Survey to consider how to make the Geological Map of the United States, and what sort of materials they are making use of.

GEOLOGICAL MAP OF THE UNITED STATES BY MR. McGEE.
— Mr. McGee's Geological Map of the United States, issued first
for the Berlin meeting of the International Congress of Geologists, October, 1885, and afterward in the "Fifth Annual Report" of the Survey, May, 1886, and finally a third time in the
"Annuaire géologique universelle, du Dr. Dagincourt," in Paris,
has a title most misleading, and especially calculated to impress
geologists with a value, which it is far from possessing. For,
instead of "exhibiting the present status of knowledge relating to
the areal distribution of geologic groups," it exhibits simply the
present status of geological knowledge of the Director of the
Survey and his associates, or at most of Mr. W. J. McGee. A
few citations will show the accuracy of the status of knowledge
possessed then by Messrs. Powell, Gilbert, and McGee, the three
responsible authors of that singularly backward geological map.

Since 1854, the Canadian River, in the Indian Territory, Texas,
and New Mexico, has been known to flow entirely over the Trias
strata, 5,000 feet thick, with mesas in the Llano Estacado area
formed by the Jurassic formation. On the geological map of
Mr. McGee, all the basin of that part of the Canadian River is
colored as entirely and exclusively Cretaceous, without even a
trace of Trias or Jurassic strata.

In 1858, an important and large geological map of Central
New Mexico was published, giving the distribution of the different formations: Carboniferous, Trias, Jura and Cretaceous. The
map compiled by Mr. McGee does not give either the Trias or
the Jura system near Santa Fé, coloring all those systems of
rocks as Cretaceous, and leaving in blank, as unknown, the whole
country about Albuquerque, the Rio Puerco and Aquafria spring.
The map pretending to exhibit the present status in 1884 is
thirty years behind our present knowledge, for all the large strip
of country between the 34th and 36th degrees of latitude, extending from Delaware Mountain (Indian Territory) to Los Angeles
and San Diego. But more, since 1854, the Triassic and Jurassic
systems have been clearly recognized and delineated in other parts
of the United States besides Texas and New Mexico; in the
map of 1884 published by the Survey they are confounded in a
single division when recognized, and when unrecognized they
are colored as Cretaceous! When the present writer called the
attention of one of the three authors of the Geological Map of the
United States of 1884 to such grave errors, the answer was sim-

ply, that it was only "a mistake in the choice of authorities." But it is for this that the Geological Survey has been created, — to give correct and exact geology, and not the erroneous specula- tions and guesses of a few individuals compromised for the last forty years by the most stupendous errors ever committed in geol- ogy. And if the three authors of the map of 1884 were unable to judge for themselves, notwithstanding one of them had ex- plored a large portion of the area, then it shows a complete incapacity on their part to deal with the subject. It was very wrong in them to make and publish a geological map of the United States, without being able to judge of the value of the work made already before.

But it is not all. Since 1845, Dr. Ebenezer Emmons has shown the existence of the Taconic system, below the Lower Silurian or Champlain system ; and Emmons, Barrande, Marcou and others have published papers and Geological maps showing that a large band of country in Eastern New York, Western Massachusetts and Vermont belongs to the so-called Cambrian of Mr. McGee. All that knowledge is ignored on the map of 1884, which pre- tends, nevertheless, to exhibit the present status of knowledge on American geology.

Another case of disregard of geological maps published is shown by Mr. McGee in the case of California, which is left a complete blank, as if its geology were then unknown, notwith- standing the publication of geological maps, covering two-thirds of that State ; one of those geological maps was issued in Paris, in 1883, and was in the hands of the Geological Survey.

The publication of such incorrect Geological Maps of the United States by the Geological Survey, is without the shadow of an excuse, for when Messrs. McGee, Gilbert and Powell con- structed their map, they had before their eyes, hung on the wall of their office, a manuscript Geological Map of the United States in nine sheets, made in 1883, and containing the actual present status of knowledge ; with a classification without confusion and with a nomenclature truly American and National, that manuscript map was generously furnished by its authors to the Geological Survey, and it has been exhibited since at the International Con- gress of Geologists at London, in 1888. where it attracted the attention of all the geologists present.*

* The map was constructed by Messrs. Jules Marcou and John B. Marcou. It can be seen in Mr. Marcou's library ; and in order to let future geologists judge what was

CONFERENCE ON MAP PUBLICATION.— The "conference on map publication" quoted previously was composed of nineteen persons, all members of the Geological Survey. Here are the principal conclusions arrived at. The time divisions "shall be called periods." Eleven periods or "geologic groups" are recognized in the United States :—

PERIOD.	COLOR.
Pleistocene . .	
Neocene . .	Orange.
Eocene	Yellow.
Cretaceous (including the Laramie)	Yellow-green.
Jura-Trias	Blue-green.
Carboniferous . .	Blue.
Devonian	Violet.
Silurian	Purple.
Cambrian	Pink.
Algonkian	Red.
Archean	

The name Pleistocene used by some European geologists and established by Lyell was adopted instead of Quaternary, because it is a "more consistent term" ; the name Quaternary being rejected "on the ground that it is a vestige of a primitive and otherwise obsolete classification and nomenclature." Very curious reasons ! in a science related to the vestiges of creation.

The Neocene period includes the Pliocene and Miocene of Lyell.

The Eocene includes the Oligocene of Beyrich ; but the Laramie subdivision is rejected in the Cretaceous, as well as the Chico group. If a great break and excellent division exists in Europe between the Tertiary and the Cretaceous, and justifies the classification arrived at by all geologists, from the time of Alexander Brongniart and George Cuvier to our present day, it is even more so in North America. The differences existing between the Laramie formation and the Cretaceous are on a greater scale and more conspicuous in America than in Europe, as well geognostically as paleontologically ; and, besides, the fine Mammalian fauna of the Laramie is identical, as far as it is possible to be at such a distance from the Paris basin, with the

Mammalian fauna discovered by M. Lemoine, near Reims, Champagne. Mr. Marsh, ignoring the existence of a quite large and numerous Mammalian fauna, in the lower part of what is incontestably the typical Eocene of the Paris basin, announced with eagerness and at the same time solemnity, that he had at last found a "mammalian fauna in the Cretaceous period;" a great *desideratum* in geology and paleontology; but the difficulty is, that the Laramie division does not belong to the Cretaceous period. So we are just at the same point that we were before the announcement of the discovery of Mr. Marsh. No Mammal fauna has been found yet in the true American or foreign Cretaceous period. For a single molar tooth found lately in 1891, in the Wealden of Hastings (England), cannot be called a Mammal fauna.

The Cretaceous period is composed in America of two great formations. So far the Geological Survey has committed the gross error of limiting the Cretaceous to one single formation composed of the following subdivisions: Fox Hill, Colorado and Dakota groups, notwithstanding that since 1854 and 1857, we know, as a part of the status of knowledge, that the American Cretaceous is well developed in Texas and the Indian Territory, with two great formations like the European Cretaceous.

The paleontologist and geologist especially in charge of the Cretaceous period for the Geological Survey, Mr. Charles A. White, unable to judge correctly of the question of classification and nomenclature, notwithstanding the published paleontological facts brought forward by Jules Marcou, and his personal acquaintance with the specimens of fossils kept in the National Museum, has declared with a certain emphasis, that "the Lower Cretaceous of Europe did not exist in North America," repeating that most inexcusable error in all his papers published by the Geological Survey until 1889; when, curiously enough, he claimed, thirty-five years after the discovery was made, recorded and accepted by all the leaders of geology, that he had discovered the Lower Cretaceous or Neocomian in Texas; taking care, as well as the Director of the Survey, to announce it as absolutely *new* and a *great discovery* due to the Geological Survey; going so far as to call it, "one of the most important events of the year in systematic geology" (*Eighth Annual Report*, Part 1, p. 82, 1889). Such is the status of knowledge in American geology existing among the officers of the United States Geological Survey. It speaks sufficiently without need of commentary.

The period called Jura-Trias is another remarkable example of the status of geological knowledge possessed by the Geological Survey. Since 1853-54 the Jura period has been recognized and described stratigraphically and paleontologically in a most beautiful typical locality in New Mexico,—the Tucumcari area. The Trias also was well defined at the same time. It is certainly curious that thirty-five years later the Geological Survey has not yet acquired enough knowledge and practical geological wisdom on the Jura and the Trias formations, in order not to unite and classify under one period only,—the Jurassic period and the Trias period.

The Carboniferous period includes the Dyas period, which is regarded only as a "subdivision sometimes called Permian." In North America the Dyas is as well developed and as distinct in every respect from the Carboniferous as in England, Saxony and Russia. It is as important and occupies a greater place and more room on the Geological Map of the United States than the Devonian period. Such a confusion on the part of the Geological Survey is anything but creditable.

The Devonian period is exact, since de Verneuil gave in 1846 the limits of its strata in the United States.

The Silurian period includes the third fauna or true Silurian and the second fauna or true Cambrian according to Sedgwick's type. The union of two periods in a single one is another example of the present status of knowledge existing among the leaders of the Geological Survey.

The Cambrian period does not correspond to the original and typical Cambrian system of Sedgwick but to the Taconic system of Emmons. The U. S. Geological Survey, instead of using the greatest discovery made in geology among the strata of the great Paleozoic epoch,—discovery made in America and by an American, is trying to destroy the national record in Geology and Paleontology. A most remarkable performance for a Survey in the pay of the National Government.

The Algonkian period, a new designation for a part of the primitive rocks, to replace the name Huronian used and coined by the Canada Geological Survey.

The Archean period, an improper name, misleading on account of the use of the designations *Archaic* and *Archaios* or

Archæo constantly employed in geology and paleontology for sedimentary deposits and fossils remains much younger than the so-called Archean period. Beside it is a confusion with the name Laurentian, used for the same period by the Canada Geological Survey.

The Algonkian also sometimes called *Eparchean* by the U. S. Geological Survey, and the Archean are only new confusions, in regard to names, introduced to designate what is well known everywhere, and will remain known always as the Primitive, or Crystalline or Azoic series of rocks.

As a whole, the "Conference on Map Publication" of the United States Geological Survey has given very meagre and backward results. If, ten years after the creation of the Survey, its officers are so far behind our knowledge of American geology in particular and of geology in general, it shows that something not only is loose in the organization, but that the base on which it rests is neither solid nor healthy.

ORGANIZATION OF THE U. S. GEOLOGICAL SURVEY.— If the Geological Survey has no plan, it has passed already through several organizations. Every year, or every two years, at most, profound changes are made ; only the leaders remain pretty nearly the same, with now and then, an addition carefully selected — not among able and good geologists — but among those that may help in keeping the machine well in hand, or even that are able politically to help the passage of the annual appropriation by Congress. It is useless to review all the organizations through which the Survey has already passed, showing at each step and change, how little both Directors, Messrs. King and Powell, were acquainted with the duties and wants of a great Geological Survey ; we shall speak only of the organization last published in the *Tenth Annual Report for 1888-89 ;* organization which has already been altered, as I am told, in several important branches.

1. Branch of Geography ; Henry Gannet, Chief Topographer. It is composed of five Divisions, beside a Division of the Irrigation Survey and a Draughting Division. The "personnel" is very numerous ; and that branch of the Survey absorbs two-thirds of the annual appropriation.

2. Mathematical Division ; R. S. Woodward, geographer in charge.

3. Division of Geologic Correlation; G. K. Gilbert, geologist in charge. In 1890, Mr. Gilbert was placed at the head of a new organization, as Chief of what is called "Geologic Branch," having in charge all the geological work of the Survey, a sort of Assistant-Director of the true Geological Survey. Before 1888, Mr. Gilbert was head of the Appalachian Division, and before 1884 he was head of the Division of the Great Basin. Those four transfers of a head of Division, in less than ten years, show in what state of anarchy, scientifically at least, the Geological Survey is constantly in.

4. Division of Archean Geology; Raphael Pumpelly, geologist in charge. Four assistant-geologists.

5. Atlantic Coast Division; N. S. Shaler, geologist in charge. Nine assistant-geologists and three volunteer aids.

6. Appalachian Division; Bailey Willis, geologist in charge. Eight assistant-geologists.

7. Lake Superior Division; C. R. Van Hise, geologist in charge. Six assistant-geologists.

8. Glacial Division; T. C. Chamberlin, geologist in charge. Four assistant-geologists.

9. Montana Division; A. C. Peale, geologist.

10. Yellowstone Park Division; Arnold Hague, geologist in charge. Three assistant-geologists and one volunteer assistant.

11. Rocky Mountain Division; S. F. Emmons, geologist in charge. Five assistant-geologists.

12. California Division; George F. Becker, geologist in charge. Three assistant-geologists.

13. Cascade Division; J. S. Diller, geologist in charge. One assistant-geologist and three laboratory aids.

14. Potomac Division; W. J. McGee, geologist in charge. Seven assistant-geologists. Under that head, geological works were carried on in Iowa, in Indiana, in the District of Columbia, in Maryland, in Central Kansas, in the Carolinas, and in the Gulf Coast (Florida, Alabama, Georgia, Tennessee, Arkansas, Louisiana, Mississippi and Texas). A rare example of an anarchical Division, which ought to be called the "*Omnibus* Division"; for Kansas, Iowa, Texas, are rather far from the Potomac river.

15. Division of Vertebrate Paleontology; O. C. Marsh, paleontologist in charge. The number of assistants and collectors of specimens is not given, but is quite large.

16. Division of Paleozoic Invertebrate Paleontology ; Charles D. Walcott, paleontologist. Seven assistant-paleontologists. It is said that since 1891 Mr. Walcott has the charge of all invertebrate paleontology in a Great Bureau, analogous to the "Geologic Branch," under the name of "Paleontologic Branch."

17. Mesozoic Division of Invertebrate Paleontology ; C. A. White, geologist in charge. The number of assistants is not given.

18. Division of Cenozoic Paleontology ; W. H. Dall, paleontologist in charge. Six assistant-paleontologists and one volunteer co-operator.

19. Division of Paleobotany ; Lester F. Ward, geologist in charge. Six assistant-paleobotanists and nine clerks.

20. Division of Fossil Insects ; S. H. Scudder, paleontologist in charge.

21. Division of Chemistry ; F. W. Clark, chief chemist. Seven assistant-chemists.

22. Division of Mining Statistics and Technology ; D. T. Day, geologist in charge. Three assistants.

23. Editorial Division ; W. A. Croffut, executive officer in charge. Two assistants.

24. Division of Illustrations ; W. H. Holmes, geologist in charge. One assistant.

25. Division of Library and Documents ; Charles C. Darwin, librarian. One assistant and several clerks and copyists.

Also a "Chief Disbursing Clerk," with several copyists.

Leaving outside of the Geological Survey all such Divisions and Branches as the Branch of Geography, the Irrigation Survey, the Ethnographic Bureau, the Mathematical, Chemistry, Mining Statistics and Technology Divisions,—all under the direction of Mr. Powell,—there remain for purely Geological and Paleontological works twenty-one Divisions, with at least one hundred and twenty-two persons employed, which number can safely be increased to the formidable one of one hundred and fifty, this being nearly the true number of persons drawing pay from the United States Treasury, for the Geological work of the Geological Map of the United States.

COST OF THE GEOLOGICAL SURVEY.— The appropriation for the year ending June 30, 1889, was $605,240. It is difficult to say how much of that sum was used for Geology and Paleontology; probably $275,000. We can say with a great degree of probability, that from 1879 to 1891 — at least two millions and a half of dollars has been expended for the construction (including printing and drawing) of the Geological Map of the United States.

This already large sum is more than the three-quarters of the whole sum required for making the Geological Map of the United States, as it is proposed in the plan submitted at the beginning of this paper; for $3,200,000 is all that is wanted, and the actual Geological Survey has already expended $2,500,000. Now let us see what has been done for that handsome sum of money.

In regard to the Geological Map of the United States, which is, or at least ought to be, the main work of the Survey, the results are preciously little and meagre ; so much so, that, at the same rate, it will be fully one century, and more, before the Geological Map will be completed; and it will require an expense of not far from forty millions of dollars! a sum ridiculously great and out of proportion to the work in view.

Some details are necessary in order that we may understand the sort of work done by the Survey, and how large sums of money are wasted.

PUBLICATION OF GEOLOGICAL MAPS. — Beside the unfortunate attempt at a Geological Map of the United States, of 1884, — pretending to give the "actual status of knowledge," when it was only a status of the knowledge of the three persons concerned, more or less, in its issue — there was an attempt to construct a geological map embracing only the States of New York, Pennsylvania and New Jersey, at the scale of 1 : 380,160. The leaders of the Geological Survey thought that it would be an easy work to transfer the geologic data to their map, requiring only *some weeks* for its achievement.

Not only some weeks have elapsed since the meeting in April 1884, in the office of the U. S. Geological Survey in Washington, of Professors James Hall, J. Peter Lesley, George H. Cook and Mr. McGee, but *years* have passed by, and we are now, in 1892, as far as ever from seeing that "Local Map," as it is called by Mr. McGee, submitted to the appreciation of American geologists.

Such schemes show better than any criticism how little Mr. McGee is acquainted with American geology, and the real "status of knowledge" of Mr. McGee and his associates. Of the 55,000 feet (about) of strata existing in those three States, 30,000 feet were wrongly classified and in a state of the utmost stratigraphic confusion ; thanks to the poor work of the three Professors consulted. And it was starting from such a base that Mr. McGee and the three Professors did try to begin the work of the construction of the Geological Map of the United States. It is to say, that these four learned geologists had the singularly unpractical idea of starting the Geological Map of our country just at the most difficult part of the task. "Unexpected difficulties have been encountered," says Mr. McGee ; and even in contending with a "Local Map," much smaller than the one intended at first, and confined to one single State, the State of New York, instead of three States, the work has remained at a standstill, and is farther from completion now than it was when undertaken by the U. S. Geological Survey.

These two failures, due mainly to Messrs. McGee, Gilbert and Powell, were easy to foresee, and are just what might have been expected by any one tolerably acquainted with American geology.

In the *Annual Reports*, we have the following geological maps :

(1). Geol. map of Ruby Hill, Eureka, Nevada, by A. Hague, 1881.

(2). Geol. map of Leadville and vicinity, Colorado, by S. F. Emmons, 1881. Reproduced, as well as the preceding map, in a large one, to be spoken of farther on.

(3). Geol. map of Virginia, Nevada, by G. F. Becker, 1881.

(4). Sketch map showing the distribution of the strata and eruptive rocks in the western part of the Plateau province, by C. E. Dutton, 1881. A good map, with an exact classification, the author having separated the Jura from the Trias, and the Dyas (Permian) from the Carboniferous.

These four maps, all included in the "*Second Annual Report*, 1880-81, were good examples, which have not since been duplicated. They give to that report a value far above all the other Annual Reports issued since.

(5). Geol. map of the Lake Superior basin, by R. D. Irving, 1882.

(6). Geol. map of Kewenaw Point, Michigan, by R. D. Irving, 1882.

(7). Geol. map of the region between the Ontonagon River, Mich., and Numakagon Lake, Wis., by R. D. Irving, 1882.

(8). Geol. map of the Porcupine Mountains, Mich., by R. D. Irving, 1882.

(9). Geol. map of the northwestern coast of Lake Superior, by R. D. Irving, 1882.

(10). Geol. map of Isle Royal, by R. D. Irving, 1882.

All those six maps are reproduced in Monograph V, ‥The Copper-bearing rocks of Lake Superior," 1883.

(11). Geol. map of Eureka District, Nevada, by A. Hague, 1882. A good map with good classification.

(12). Preliminary geol. map of the Northwest. by R. D. Irving, 1884.

(13). Geol. map of northwestern New Mexico, by C E. Dutton, 1885. A good map with good classification.

(14). Geol. map of Driftless region and environs, by T. C. Chamberlin and R. D. Salisbury, 1885.

(15). Geol. map of Martha's Vineyard, by N. S. Shaler, 1886.

(16). Geol. map of Central Wisconsin, by R. D. Irving, 1886.

(17). Geol. map of Northeastern Minnesota, by R. D. Irving. 1886.

(18). Geol. map of Lassen peak (California), by J. S. Diller, 1887.

(19). Geol. map of portions of Ohio and Indiana, by E. Orton, 1887.

(20). Geol. map of Mount Desert Island, Maine. by N. S. Shaler, 1887.

(21). Geol. map of a portion of northwestern Colorado, by C. A. White, 1888.

(22). Geol. map of the Northwest, by Irving and Van His, 1889.

In résumé, only four good geological maps, by A. Hague, S. F. Emmons and C. II. Dutton, of small districts in New Mexico, Colorado, Utah and Nevada. For a period of ten years, it is a very small showing.

The great 4to *Monographs* issued by the Survey, give reprints of almost all the geological maps contained in the *Annual Reports*, with additions of some maps in the districts of Leadville, Comstock lode and the quicksilver deposits of California. All the geological maps of the Leadville district by S. F. Emmons are excellent and reflect great credit on their author.

As to the geological maps in the Atlas, accompanying the Monograph on the Geology of the Quicksilver Deposits, by George F. Becker, they are all based on wrong classification and erroneous chronologic age of the strata, and should be worked anew by competent geologists before being used in the preparation of the Geological Map of the United States.

During the meeting of the International Congress of geologists, at Washington, a geological map of the vicinity of Washington, at the scale of 1 : 62,500, marked "Preliminary edition," 1891, was distributed with a "Guide to Washington."

With a few insignificant geological sketch maps issued in some *Bulletins* and in the *Annual Reports*, we have, above, all the geological maps issued by the U. S. Geological Survey up to 1892 ; during a period of twelve years.

PUBLICATION OF REPORTS AND PAPERS IN THE "ANNUAL REPORTS," THE "BULLETINS" AND THE "MONOGRAPHS." — The publications of the Geological Survey comprise a few good memoirs and papers drowned among papers either of a very inferior quality or even of no value whatever for geological purposes. Without speaking of papers on Topography, Geodesy, Chemistry, Mining, etc., which ought not to obstruct the publications of the Survey, there are already too many papers which ought never to have been accepted for publication, either because they are not geological at all, or because they are bad papers. I shall confine my remarks to a few.

In the last *Annual Report* we have a very long paper with many illustrations on "Swamps." Such a paper may be interesting in a Forestry report or an Agricultural report, but for the construction of the Geological Map of the United States, it is absolutely useless. The geology in it is insignificant, even with the formation of morasses. Botanists and timber men may go to the "Dismal Swamp" district of Virginia and North Carolina to make observations on pine and cypress trees or on morasses and swamps of all kinds ; but it is not the work of a geologist. Physical geography may note the existence of swamps ; but all that sort of work ought to be left to the topographers, the Coast Survey engineers, the lumber surveyors, etc.

In the Bulletins, one marked No. 57, "a geological reconnaissance in southwestern Kansas," is so poor in regard to the age of strata, their classification and nomenclature, that it ought to have been kept out and placed aside with the report on the geology of Texas, spoken of in the *Eighth Annual Report*, p. 172, 1886–87, recognized, at last, as unfit for publication.

"The Geological Exploration of the Fortieth Parallel" has given, with sufficient details, the description of the Comstock lode at Virginia City and its vicinity. Just as if that already expensive work issued by the United States government did not exist, a new survey has been made by the Geological Survey and two *Monographs* published at very great expense, which are of no use to any body, except the very rich mining companies holding the Comstock lode. No new scientific facts have been added by the publication of *Monograph* III, "Geology of the Comstock lode and the Washoe district," nor *Monograph* IV, "Comstock Mining and Miners"; it is only a new bounty of public money given to millionaires. The new survey has not the shadow of an excuse; it is simply lobby work for the benefit of silver magnates.

Looking over the list of publications issued by the Survey, it is surprising to find so many papers on surface geology. Their number is already in complete disproportion to the importance of the Pleistocene period, the most easy to study of all the geological periods and one which ought to attract the least attention. To observe old moraines, kamers, lake terraces, old river beds, etc., etc., is so simple, that it requires almost no knowledge of geology, being only geographical work. Any student, with a little industry and perseverance, will trace the shores of old lakes, the moraines, the glacial deposits, old river beds, etc. Such work ought to be left to volunteers and free geologists, not in the pay of the United States government. Nothing shows more plainly the total absence of plan and scientific control than those well-nigh endless papers on old lakes, glacial geology, morasses, sand beaches, old river courses, etc.

Another evil and useless expense in the publications of the Survey is incurred not only in the repetition of papers in the *Monographs*, already published for the most part in the *Annual Reports*, but also in the length given to certain subjects quite unwarranted by their true scientific value. For instance, all the works of Mr. R. D. Irving on Lake Superior have been first published in the *Annual Reports* and then reprinted as a great 4to. *Monograph*, without sufficient additions to justify the expense.

" The Geology of Nantucket," *Bulletin* No. 53, with 10 plates and in 55 pages, would have been much improved if reduced to one plate and only six pages. In such a reduced form the sand of Nantucket will have received all the attention it can command scientifically. The " Report on the Geology of Martha's Vineyard," in the *Seventh Annual Report*, occupies 62 pages, with 11 plates and 9 diagrams; yet even with this luxuriance of plates, not a single one of the truly interesting specimens of fossils existing in the Island is given. Gay Head is the only locality there worth surveying stratigraphically and paleontologically; all the rest of the island may be disposed of in six pages. The author, on the contrary, expends all his theoretical knowledge on ground moraines, frontal moraines, kames, terrace drift, erosion, etc., and leaves the paleontology entirely out, without a single fossil figured, described or even named; and he is Professor of Paleontology at one of our great universities! All the 62 pages of the author might have been concentrated with much advantage into ten pages, and the illustrations reduced to one map and two plates of views. The paleontology, comprising fossil invertebrates, fossil crustacea, fossil vertebrates and fossil plants, ought to have been added, as the most important and only interesting feature of the survey of Martha's Vineyard. It is a new example of the total absence of scientific control in the Geological Survey. The papers quoted above are not the only ones; they have been taken at random and might be triplicated most easily.

If the geological maps and the geological descriptions issued by the Survey are — with a few exceptions — not of the standard which may allow them to be compared to their advantage with similar works published in England, France, Germany, Scandinavia, Russia, etc., — let us turn to the Paleontological publications. Up to the present time the Geological Survey has not yet published a single good figure, well drawn and well printed, or useful for reference. The " Division of Illustrations " is very efficient in regard to views and photographic works of scenery; but for fossil drawing it is a complete failure; and foreign paleontologists who are accustomed to good figures of fossils and know how generous our Government is in its expenditure for the Survey are quite astounded by the worthless characters of all the fossil plates. They do not understand it. There is always the same reason,— a total absence of scientific control in the Survey.

The paleontologic publications of the Survey are generally, with a few exceptions, second order works, and too often poor and unworthy papers. Some ought never to have been accepted for publication. For instance, in the *Fourth Annual Report*, 1883, there is "a review of the fossil Ostreidæ of North America," most incorrect in every respect, zoologically as well as stratigraphically and bibliographically, and with most defective and deceitful figures. The author placed in the Cretaceous period Ostracæ belonging to the Jurassic and Tertiary Eocene periods, confounding species so different that any young student would have had no difficulty in distinguishing them. The same author has succeeded in throwing into the greatest confusion the "California Division" of the Survey by making a perfect conglomerate of errors with the genus *Aucella*; not to speak of his other gross errors on the Tertiary fauna of California, the Triassic fauna of Texas, the Jurassic fauna of the Tucumcari area, and the Cretaceous fauna of Texas, New Mexico and Kansas.

Two *Monographs* on fossil fishes are so far behind our present knowledge that it is difficult to understand the choice of the Director of the Geological Survey for the work, when we possess in America such excellent paleontologists for fishes as Messrs. E. D. Cope and Orestes St. John.

The *Monographs* of fossil plants from the Mesozoic of the Atlantic shores in Virginia and Maryland ought to have been carefully revised, as well for the determination and description of species and for geological references as for figures and drawings, before being accepted for publication.

EXTRAVAGANCE IN THE MANAGEMENT OF PUBLICATIONS. — All the publications of the Geological Survey are issued in two editions; the first comprising the usual number (always very large) of documents ordered by Congress, which is already fully sufficient for all the public libraries existing in all the States and Territories; and the second edition for the special use of the Geological Survey. This second edition is enormous, and out of all proportion to the real want, for the number of *Annual Reports* is 5,000 copies for each Report, of the *Bulletins* 3,000 copies, and of the *Monographs* also 3,000 copies.

I shall refer only to the second edition for the use of the Survey, although the first edition for Congress and the general use of the Government is already so large and so generously distributed, that many copies find their way into the hands of geologists and public libraries.

Any geologist well acquainted with the present status of geology all the world over will say that an edition, for the use of the Survey and for sale, of 1,000 copies each, of the Bulletin, Monograph and Annual Report, would be amply sufficient and above what is done in any other country where a Geological Survey exists, such as England, France, Switzerland, Germany, Austria-Hungary, Scandinavia, Italy, Spain, Russia, India, etc. Not one of these countries issues 1,000 copies of any of its publications, some being limited to 300, others to 500 ; very few go as far as 600 copies.

But in order to be in excess of all demand and to satisfy all just desire of procuring the publications of the Survey for serious use, let us name an edition of 1.200 copies as a *maximum*, which ought never to be surpassed on any plea, even for the reason so rashly put forward by Director Powell in his answer to the just criticism of Mr. Alexander Agassiz, before the Congressional Commission of 1884-85.

Mr. Agassiz had said that : "The method of publication, the mode of distribution, the size of the editions are wasteful and extravagant. Editions running into the thousands and tens of thousands are often published (by the various bureaux), and there are not 500 people or institutions whom they will do any good, or to whom such purely scientific publications can be of any service." Mr. Powell's answer is characteristic of the scientist-politician, for he says : "There is a sentiment . . . . that would exclude the people at large from a knowledge of the progress of scientific research. In the lands where this sentiment has its home it is desired to establish in emulation of the hereditary aristocracies scientific aristocracies, which, it is claimed, should also have official recognition. In Europe this sentiment has lent its aid to the policy of publishing very small editions of great scientific works, but the policy has not been generally adopted in America, because the sentiment has gained no foothold in this land of free institutions. . . . . . The Director of the Geological Survey has long been of the opinion that the time will soon come when all of the scientific publications of the General Government. including those of the Geological Survey . . . . will be distributed to one or more public libraries in every county of the United States. *This is imperatively demanded, in order to secure a wise and just distribution among the people of the sources of advancing knowledge.*" No one but a scientific demagogue could have advanced such statements and in such terms,—base flattery to the people and a total absence of regard for truth. Such language, until now, was never heard in Science, but has been confined to political clubs.

A few exact enumerations will dispose of all these "sentiments" and the scientifico-political flourish of the Director of the Geological Survey.

The number of geologists everywhere is, and will be always, very small. In North America it is smaller than anywhere else, except in Japan. In a population of about 70 millions (the Dominion of Canada and Newfoundland included), the number of geologists is below five hundred. The Geological Society of America numbers 200 fellows, who comprise the bulk of really active geologists; if we say five hundred geologists in North America, as a grand total, we are certainly a little in excess of the true number. In England the number of geologists is about 2,500 in a population of 38 millions, and the Geological Society of London has about 1,300 fellows. In France the number of geologists is about 1,400 in a population of 38 millions, and the Geological Society of France has about 600 fellows. In Switzerland the number of geologists is about 100 in a population of nearly 3 millions. In Italy the number of geologists is about 350 in a population of 30 millions. In Germany the number of geologists is about 1,600 in a population of 48 millions.

Proportionally the number of geologists in North America, and more especially in the United States, is far below the average number in European countries. Switzerland has been a republican country for above six centuries, and is a "land of free institutions," without "hereditary aristocracies." France is also a republican country and is also certainly a "land of free institutions." And although both countries contain many more geologists than the United States, their publications are neither wasteful nor extravagant as regards methods of publication, mode of distribution, sale to the public, or the size of editions. And certainly the science of geology is in a more healthy condition in every respect in Switzerland and France than in the United States. In both countries they have a Geological Survey, with a board of Regents, called a Commission or Council, in which are placed all their best geologists, representing equally the different opinions on all questions purely geologic.

To suppose that the publications of the Geological Survey ought to be found in one or more public libraries in every county in the United States is the most erroneous notion imaginable. If so placed, nine hundred and ninety-five in one thousand would never be of any use to anybody; they would be wasted copies, occupying room which is sure to be wanted for other works more in demand and truly useful.

Let us take one work issued by the Geological Survey and see the number of persons to whom it may be really of use. And in order to be liberal, I should choose the "Annual Report," issued at 5,000 copies for the special edition of the Survey. The last one, called the *Tenth Annual Report*, contains: 1st. The Report of the Director, which is an administrative short résumé of the work in progress, with an unfortunate and entirely valueless essay, with diagrams for standard colors, upon the representation of formations, "fossiliferous clastic rocks," superficial deposits, ancient crystalline and volcanic rocks. 2nd. Administrative reports of the different "Heads of Divisions",— which interest nobody but their writers. 3d. "General account of the fresh-water morasses of the United States, with a description of the Dismal Swamp district of Virginia and North Carolina"; a paper out of place and without the smallest interest to geologists. 4th. "The Penokee iron-bearing series of Michigan and Wisconsin," with twenty plates of thin sections of rocks and ore deposits; a lithologic paper which may interest about fifty persons in North America, and two dozen at most in Europe, Asia and Australia. 5th. "The fauna of the Lower Cambrian or Olenellus zone," a paper which contradicts two previous memoirs issued in the *Bulletins* by the same author, and establishes a *chassé-croisé*, doubly erroneous, and absolutely unique in stratigraphy. If there was the faintest scientific control in the Survey, that paper would have been submitted to experts before being accepted for publication; for it is the most worthless production yet issued on the Primordial fauna of the world, as well stratigraphically, synchronism and classification, as paleontologically. The number of persons in North America interested in the paper is extremely limited,— about two dozen at most, and outside of America eighteen others may look over it,

So, in all, the *Tenth Annual Report* interests 74 American geologists and 42 foreign geologists, a total of 116 persons. And for that small number five thousand (!) copies have been printed besides the already large amount of copies for the use of Congress and the General Government. When I say, that twelve hundred copies of the "Annual Report" is amply sufficient to all present and future reasonable requisite; it may be regarded almost as an extravagant estimate by Mr. Alexander Agassiz and other good judges of publication of scientific work, but I want "to secure a wise and just distribution among the people of the sources of advancing knowledge", according to Mr. Powell's phraseology.

The cost of publications during ten years may be recorded as $350,000, of which $250,000 might easily have been spared with advantage not only to the public purse, but also to the Survey, for the number of volumes is accumulating so fast, that a whole building is required in which to store copies of books undistributed; and the want of place, one day or another, will be such that the number of copies printed will either be cut down to a much smaller number, or the volumes undistributed sent back to the paper mill to be converted into pulp and new paste.

PALEONTOLOGICAL WORKS.—That the first explorations west of the one hundredth meridian,—when all the immense regions of the Plains, of the Rocky Mountains, of the Parks, of the Great Basin, of the Sierra Nevada, of the Columbia river, of the Colorado river, of New Mexico and of Texas, were a total wilderness, difficult and dangerous to travel,— should publish everything of Natural History collected, was proper and extremely useful for the progress of our knowledge of almost half of North America. But now it is very different.

The end to be attained is the construction of the Geological Map of the United States, and everything not absolutely necessary for that purpose should be left to private individuals, specialists, natural history societies, academies of science, universities, colleges, schools or even States. For instance, Botany and Zoology are abolished from the program of the Geological Survey: but it is not enough, we must go further and suppress all branches not actually needed for the preparation of the great Geological Map.

For instance, vertebrate paleontology and mining surveys are such special studies, that they ought to be left to special investi gators, or to learned societies and academies, or to industrial corporations. The fossil vertebrata are always very rare, difficult to collect, and cannot be used practically, when in the fields, as characteristic fossils to find the age of strata. As to mining surveys, they are useful only to Mining companies, each one of which knows perfectly well, within a few thousand dollars, the exact value of its properties; and it is preposterous for the government of the United States to prepare and publish very costly memoirs for the benefit of millionaires who are already harassing Congress to increase the value of their mines, by taxing all the citizens by protective and most unjust and injudicious duties; some even being prohibitive, as is the case with tin, copper, coal and certain silver ore.

The vertebrate paleontology is a choice example of the bad influence exerted by the help of the Geological Survey, which, instead of aiding in the progress of that branch of paleontology, has caused a true public scandal, without any just reason, and simply by the favor conferred most injudiciously upon one scientist in preference to any other. Without entering into the merit or demerit of the paleontologists engaged in the serious and almost disgraceful polemic and controversy, which was made public in January 1890, by publications in the "New York Herald," and other newspapers, it is impossible to pass it over.

Before the law creating the office of Director of the Geological Survey was enacted, in March 1879, Mr. O. C. Marsh, as acting President of the National Academy of Science, took a very active part in the question, showing partisan views and very incorrect notions in regard to Government surveys and other matters entirely out of his line — the study of the vertebrate. If any one was prevented by his interference before Congress from getting an office in the Geological Survey, it was Mr. Marsh. But as soon as the Survey was well under way, in 1881, Mr. Marsh received an appointment, as "Head of Division of Vertebrate Paleontology," with more of the people's money at his disposal than any other branch of the Geological Survey, except the Geographical Division. It certainly looks like a division of spoils applied to the Geological Survey. And this is not the only case, for, in 1884, when a change in the Presidency of the United States and the House of Representatives from republican to democratic ascendancy was made, a political nomination, with creation of a new Division, entirely uncalled-for and needless, was made, in order to influence the Committee on appropriations, and also at the same time to prevent the nomination of a new Director of the Geological Survey.

The creation of a Division of Vertebrate Paleontology and the choice of Professor Marsh were very injudicious and reveal incompetent direction on the part of the leaders of the Geological Survey, so far as the interest of the Geological Map of the United States is concerned, as also a judicious expenditure of public money. Let us notice the purely scientific objection to that creation and nomination. The study of fossil vertebrata is not confined to a single paleontologist. Other paleontologists have studied — and some with great success — the fossil vertebrata of North America; and to give the powerful patronage of the Geological Survey to one scientist only, with a large annuity of the fund at the disposal of the Survey and resources from which to draw for exploration, travelling and express expenses, is to give to him a powerful and dominating position, to the great disadvantage of other paleontologists, and the placing of a formidable bar in the way of fair emulation. Practically, Mr. Marsh is placed in such a favorable position — outside of his private resources, which are already very great (being a Professor at Yale University and a millionaire)— that all other vertebrata paleontologists have greater difficulty in getting good specimens or in preparing and publishing the results of their researches. We all know that a certain, we may say healthful, rivalry and emulation will always exist among workers in the same field of research, in the endeavor to obtain, describe and show great and rare collections; and it is best and more just to let the field be entirely free, without the interference of the United States Government.

Professors and Curators of museums belonging to universities, scientific societies and great cities, are well prepared and well equipped to make special researches; and we have already in our country—and we shall have more by and by — establishments of public instruction, which are the proper places for the study and collecting of fossil vertebrata. It is neither just nor a good policy to crowd out other establishments and scientists by the conferring of such privileges as are granted by the General Government of the United States to Mr. Marsh and to Yale College. It is very detrimental to the progress of vertebrate paleontology.

If the Geological Survey, instead of giving every year the large sum devoted to the Division of vertebrate paleontology to only Messrs. Marsh and Newberry, had divided it equally among all the vertebrate paleontologists of the United States, it would have been simple justice to such able and good observers as Messrs. Cope, Scott, Osborn, Baur and St. John are.

But by far the best solution is to suppress entirely the Division of Vertebrate Paleontology, and to let that field of investigation be open and entirely free from all interference by the General Government of the United States. Specimens of fossil vertebrates, collected accidentally by the officers of the U. S. Geological Survey, in the regular course of their researches, should be placed in the National Museum at Washington, so as to be easily accessible to all special students of those branches of paleontology, as is done at the South Kensington Museum in London, at the Jardin des Plantes in Paris, etc.

Paleontological work must be confined in the Geological Survey to invertebrates and fossil plants, and even in those branches, to the most characteristic and truly useful in practical geology. Each geologic period may be sufficiently represented on fifty 4to plates, more or less, of very good figures, well drawn and well printed; and the paleontological volume of each period might be joined to the geological volume as a supplement.

"Personnel." of the Geological Survey.—It is always a delicate task and a very difficult one to speak of persons ; and when those persons are scientists it is more so. for no class of men are more sensitive or quick to take offense. Louis Agassiz, at the end of his life, after his long experience in Europe and in America, used to say, that scientists are the most difficult class in society to get along with, and a very thorny set of men.

However, the officers of the United States Geological Survey are public men, paid from the people's Treasury, and they cannot escape the responsibilities they incur before the people. Since men competent to judge of these matters and willing to express their opinions are extremely rare, a greater responsibility and a higher duty devolves upon them to speak frankly, honestly, and justly. For they alone can see injudicious expenditure of the public money and the incompetency of direction in scientific work.

Taking only the Director, Heads of Bureaux or Branches, and a few Heads of Divisions, we shall consult the last list published in the *Tenth Annual Report*. Some changes have been made since in that list, which is dated 1888–89, but not any important enough to alter in any way our estimate and conclusions. Leaving out of consideration the "Branch of Geography," as well as the Divsions of "Vertebrate Fossils," of "Chemistry," of "Mining Statistics and Technology" (which ought to be separated from the Geological Survey and transferred to other Government Bureaux, or even suppressed), we shall review the scientists who have in their hands the direction of the work for the construction of the Geological Map of the United States.

The Director Mr. Powell may be consigned with advantage to the Bureau of Ethnology, with a military pension beside if necessary, should he not already have one. It will be an economic investment and a good policy on the part of the Government to allow him a good salary and dispense with his services in any work requiring geologic and paleontologic training and knowledge.

Mr. Gilbert is devoted to the Pleistocene period and will make a good Head of that Division. But to place into his hands all the "Geologic Branch" of the Survey is a great error of judgment. He does not know practically, how to make synchronism or correlation, how to make sound classification, nor what nomenclatures are. He does not know fossil remains, and is not even well acquainted with the history of the progress of geology in America. He has never studied any typical locality, or any geological period in Europe. Besides, his experience in the construction of geological maps is extremely limited, and outside of superficial or surface geology he is incompetent to undertake or direct any stratigraphical and geological work.

Mr. McGee is a very hard worker, and is well qualified for bibliographical researches; he knows just enough geology to make an excellent Librarian of the Survey; and his tastes for Catalogue, Dictionary, and *Thesaurus* of all that relates to Geological Science can be turned into good use as Chief of the great library of the Geological Survey. Mr. McGee does not possess any of the qualities or the knowledge necessary to direct the researches on the synchronism and the classification of periods and . formations, because he is not a paleontologist — or at least does not know enough of practical paleontology to make use of that science —, and he has never studied the typical localities of

Europe. It is injudicious to trust him with the construction of the Geological Map of the United States, or any part of it, for he has no familiarity whatever with nine-tenths of the series of American strata. His essays of two geological maps, although they were compilations, were so unfortunate, that it is evidently out of the question to put into his hands any work requiring a profound knowledge of geology or paleontology, or discrimination in the use of authorities.

Mr. McGee is truly a lover of books, but too much inclines to theorise and to coin new names; and, if it depended on him, he would soon change the whole language of geology and make it with words rather curious. All the numerous works of the different Geological Surveys published in the English language in England, Canada, Asia, Australia, Africa, Newfoundland and the West Indies, are written in good and intelligible English, resembling somewhat the fine writings of Lyell, Sedgwick, Conybeare, De la Beche, John Morris, Edward Forbes, Jukes, Ramsay, Prestwich and Archibald Geikie. But with the publication of the U. S. Geological Survey it is otherwise; new names are coined by the dozen, others are all terminated in *ic*, and others hardly used at all before in geology are employed constantly with significations rather exaggerated and very odd.

I shall quote a few of them taken from the Director's reports, and the papers and reports of Mr. McGee. In the *Fifth Annual Report*, we read: Diastrophic, Hydric, Glacic, Eolic, Biotic, Anthropic, Lithic, Petromorphic, Geochronic, Choric, Geomorphic. In other reports and publications we read: Taxonomic, Eparchean, geomorphology, antecedent valleys, consequent valleys, superimposed valleys, epigenetic, autogenous, inequipotential, paleobotany, homogeny, irostacy, kaineontologic, kaineontologist, etc., etc. The curious part of it is that Director Powell is anxious to secure to the people "the sources of advancing knowledge," and the first thing he does, in collaboration with some of his favorite assistants, is to write an English which will not be understood, not only by the people in general, but even by the learned; issuing and forging new names right and left, without the smallest necessity, and too often against the rules of philology.

Mr. Walcott is an excellent collector of fossils for he possesses a rare talent for finding fossils and fossiliferous localities.

In paleontology his work is very unequal, sometimes being acceptable at other times erratic and unreliable. Not having received
a zoological education, he is only a second rate paleontologist.
Some of his paleontological papers may be classified as having
some real value;[1] but he has not published anything to compare with the paleontological memoirs of Barrande, Salter,
Angelin, Linnarson, Brögger, Holm, Matthew, Davidson and
Barrois. As to stratigraphy and classification, Mr. Walcott is
anything but a successful and exact observer; and in order to
throw dust in the eyes of those who know little or nothing of
the question of the Lower Paleozoic rocks, he has made invariably
in all his papers a great show of his "principles." He pretends
that he always acts according to the best methods, the most exact
observations, hammer in hand, and he gives all his conclusions
with "considerable confidence." In fact never was a geologist
and paleontologist with such an amount of conceit.

Seldom has a geologist begun public life under more favorable auspices and at a more propitious time. Calling himself
the favorite pupil of honest Colonel Jewett, it was expected that
he was a man who could be trusted as an exact observer and as
true to his words, his expressed opinions and his promises. But
in less than six months he lost the sympathies and even the respect of all those who place sterling scientific honesty and truth
in geology above all other accomplishments.

The study of the rocks containing the three faunas called infra-
primordial, primordial and supra-primordial or Taconic period
has created such bad feeling and produced such irreconcilable
divergence of opinions, that the appointment of foreign geologists has become a necessity, in order to arrive at a proper classification and good descriptions. The objections often raised
against foreigners will be here out of place; besides, we have
precedent: Hassler, the first Director of the Coast Survey, was
a Swiss from Aarau; the best military engineer and constructor
of our old sea-coast fortifications was Simon Bernard of Dole
(Jura), a French general, who was secured by our Government

[1] As an assistant of Mr. Hague in the Survey of the Eureka district, Mr. Walcott
was very successful in collecting rare paleozoic fossils, and he gave a valuable Monograph of the "Paleontology of the Eureka district," 1884. His success was mainly
due to his association with Mr. Hague, a good stratigraphist; but as soon as he directed himself, and made stratigraphic work of his own, he has fallen into the greatest errors.

at the disbandment of the French army in 1815, and was appointed Brigadier General in the U. S. Engineer Corps; Agassiz was a Swiss, and yet he was appointed a Professor at Harvard University and a Regent of the Smithsonian Institution.

There are now four foreign geologists particularly well trained and entirely qualified to make a good survey of our oldest paleozoic rocks. One resides at St. John, New Brunswick, Mr. G. F. Matthew; the second is at Lille, in France, Mr. Charles Barrois; the third is at Christiania, Norway, Mr. W. C. Broegger; and the fourth is at Stockholm, Mr. G. Holm. At least two of these four gentlemen ought to be secured by the Geological Survey, giving them handsome salaries and proper situations in the Survey. Some of our young American geologists, who have already shown by previous work ability, might be chosen and attached as assistants to the foreign geologists. Certainly, Mr. Edward O. Ulrich of Newport, Kentucky, would make an excellent assisant; and a few others may easily be picked up. The work to be done is too important, too difficult, and has been too long in the hands of incompetent stratigraphists and paleontologists, to be continued any longer on the line adopted by the Geological Survey. A change of base is an absolute necessity in order to arrive at the truth.

Of the four actual leaders of the Geological Survey not one is the right man in the right place, as I have shown with sufficient proofs.

Of the others, "Heads of Divisions," some are good, and several of the assistants are also able and exact observers. I shall signalize two Heads of Divisions manifestly unfitted by their knowledge, or, more exactly, absence of knowledge, for the special work assigned to them. One, Mr. C. A. White, is a very unfortunate choice for all geognostical and paleontological work on the Mesozoic rocks and fossils. While living in Burlington, Iowa, as a business man, he collected and studied fossils. He has some knowledge of the Carboniferous period in Iowa, but is entirely deficient in studies regarding the Cretaceous, the Jura, the Trias and the Dyas. He does not know any typical locality in the old world, and his biological studies are of a very light order. His taste is for description of fossils, and he is tolerably well posted as regards Lower Carboniferous fossils, — as a third rate paleontologist. As a stratigraphist and classificator he has contantly blundered, and cannot be relied upon for any exact observation.

As to the "Head of Division of Library and Documents," Mr. C. C. Darwin, he is a complete stranger to the sciences of geology and paleontology. To handle a special library devoted exclusively to one great science requires a profound knowledge of it; and for the Geological Survey a good geologist for librarian is a necessity.

At first Mr. Darwin thought that he was able to undertake bibliographical researches and publish complete catalogues on all branches of American geology. It was simply a scheme without a base to set upon, for Mr. Darwin is not in a condition to undertake with success catalogues of papers or of geological maps or of references to American geology, many of which are scattered in many publications and in places where no one will look for them except able geologists. It will be best for all concerned, the Geological Survey as well as the National Library, if Mr. Darwin is sent back to his old position at the National Library.

Only a few other general remarks will be sufficient to show some of the very weak points of the actual "personnel." As a rule, no one ought to be appointed on the Survey who cannot give all his time to his work. To draw at the same time several salaries may be very pleasant, but it is not consistent with good and steady work. If we look at the list of the "personnel" on the pay roll, we find that almost half of them are professors in universities, colleges, etc. The Geological Survey is considered by some as an excellent occasion to increase their salaries and to direct a field summer school, and every year during the vacation parties start in every direction, assisted and accompanied by students and friends. Such vacation works are always very superficial and of a low standard. It is impossible for a geologist to attend to his duties in a university, college, school, etc., and at the same time to carry out important work for the Geological Survey; one or the other is neglected. Generally the Geological Survey work is the one which suffers the most, because he knows by experience that he can neglect it with impunity, every "Head of Division" in the Survey doing what he pleases and when he pleases, and that his name with his title of Professor at Harvard, Yale, Columbia, Cornell, John Hopkins, Amherst, etc., is, above all, what is wanted by the Director of the Survey. To be sure, it will be a bad policy to exclude absolutely from the Geological Survey the use of Geologists and Paleontologists already in other official positions. It may present some rare cases when a specialist's advice, views and observations on some difficult and embarassing points or questions may be required. But the appointment then must be temporary and limited to the special questions; and as soon as the work of the expert is finished he ought to be dismissed, and not retained as "Head of Division," or even as Assistant in the United States Geological Survey.

PRESENT CONDITION OF GEOLOGY IN THE UNITED STATES. — Until 1844 American geology was in a very healthy condition. The leaders were all excellent observers, good classificators, just, honest and gentlemen. Dr. E. Emmons was the first assailed and subjected to bad treatment and most unjust indignities, because he made the discovery of the Primordial fauna and a great system of strata below the Potsdam Sandstone. It was the beginning of acts most reprehensible, made or directed by an incompetent geologist, formerly the assistant of Emmons and afterward his colleague in the New York Geological Survey. Lyell, de Verneuil and Marcou became in succession the objects of the morbid jealousy of that person, who finally succeeded in creating an association with the special aim to arrest, or at least to retard by all means the progress of American geology. Incompetency in practical geology, classification, nomenclature, united with incorrect determinations of fossil remains and a complete want of knowledge of typical localities of all the geological periods, has led the Paleontologist of New York and his associates to undertake an impossible task, far above all their capacities united. And they have gone constantly astray, oscillating among errors, in all their attempts to monopolize for themselves the geology of North America. All their effort and essay at synchronism, equivalency of formations, classification of strata, have ended so disastrously that it is sufficient to enunciate them in order to show their failure. 1st. No fossiliferous rocks exist below the Potsdam sandstone. 2d. The so-called Primordial fauna, instead of being below the second fauna, is above it. 3d. All the Taconic system belongs to the Lower Silurian (original Cambrian), the Upper Silurian, the Devonian and the Carboniferous metamorphised by heat. 4th. The Dyas system (Permian) does not exist anywhere in America. 5th. The Triassic system of Texas, New Mexico, and the Indian Territory belongs to the Dakota group of the Cretaceous. 6th. The Jurassic system of the Tucumcari area and of New Mexico belongs to the Dakota group and is "unquestionably" Cretaceous. 7th. The Neocomian or Lower Cretaceous does not exist in North America for "it is a well known fact that we have in North America no strata which are, according to European standards, equivalent with the Lower Cretaceous of Europe." 8th. The Eocene of California (called Chico and Tejon group) belongs to the Cretaceous period. 9th. The Eocene of Laramie belongs to the Cretaceous. 10th. The auriferous gravels of California, typical of the Quaternary period, belongs to the Tertiary. 11th. The ice age does not exist and its pretended discovery is due to vagaries. 12th. There are no living glaciers on the Mount Shasta, nor in the Sierra Nevada of California, etc. In paleontology the errors forced on American fossils by the Paleontologist of New York and his associates are as conspicuous and numerous as those in the classification of strata, nomenclature and equivalency.

The duties of the United States Geological Survey were plain from the start. Bring back American geology to a healthy position as it was before 1845 ; repair as much as possible the injustices done to exact and good observers; make use of all the discoveries and good works published since 1845, and refer them to their true authors. But an incompetent direction, on the contrary, has given, since 1879, a renewal to all the bad proceedings used by the despotic association inaugurated between 1845 and 1854 ; and the Geological Survey, instead of fulfilling its mission of inaugurating an era of justice, has gone astray, endorsing almost all the ridiculous errors saddled on American geology, and it does all it can to cover unexcusable blunders, dishonest practice and incredible mistakes.

It is not all, for when badly employed a great Government institution and Bureau, can do more harm than the most mischievous association. The Geological Survey enrolls systematically all the geologists having positions in the leading universities, colleges, etc., in order to control all the observers, which may make a serious opposition and embarass the passage of large appropriations by Congress. Already a great number are on its list, and every year it is increased, until it will embrace all the most active and influential geologists in the country. Of course, there are a few opponents; but they are silenced as much as possible, by special manœuvres, recalling political practice.

For instance, the Geological Survey captures, one after another, all scientific organizations, Societies and Academies. The Geological Society of America is virtually under the Geological Survey control; and any paper presented which displeases any one of the Geological Survey is simply returned to its author, under the easy plea of want of space. The National Academy of Science is also controlled, in regard to geology, by persons at the devotion of the Geological Survey, or even by officers of the Survey; and cannot be relied upon for anything relating to reforms, or a competent direction, or even for good advice. All the scientific societies of Washington, or having meetings in Washington, are controlled by the Geological Survey, at least for all subjects relating to geology. The Academy of Science of New York, is also in the hands of the Geological Survey. *The American Journal of Science* of New Haven is entirely devoted to the Geological Survey.

The practice of letting members of the Geological Survey publish observations made when on duty, is very prejudicial to the Survey, and also to American geology in general. It often brings out papers rather discreditable, being either erroneous or indigested; and it would be best to reserve all the facts to be co-ordinated at the end, when the whole subject should be well in hand. The Survey is a very serious undertaking, it is national, and ought to act always with dignity and openly; and it is regretable to see papers written on partisan views declared "unfinished" by the author[1], and very incomplete. No good can come out of such publications either for the Survey, or the progress of American geology.

THE INTERNATIONAL CONGRESS OF GEOLOGISTS AT WASHINGTON. — A remarkable instance of the proceedings used by the Director of the U. S. Geological Survey, is the capture of the International Congress of Geologists. At the meeting of the International Congress of Geologists at London, an invitation was extended for its next meeting, by the city of Philadelphia, endorsed by all the American geologists present at the meeting of

1 *Correlation papers—Cambrian* by C. D. Walcott, p. 17. "This report is an unfinished memoir." Washington, 1891.

the 21st of September, 1888; and it was accepted, voted by accla-
mation, and the President Prestwich addressed a cable telegram
of thanks and acceptance to the Mayor of Philadelphia.   The invitation was most pressing and liberal, and was made by the
Mayor, the Aldermen, the Council, the University of Pennsylvania, the Courts, the Scientific Societies, the Directors of Banks
and Railroad corporations, etc.   (*Compte Rendu de la
Quatrième session du Congrès géologique international*,
Londres, *1888*, *pp. 45-50.*)

Director Powell is not the man to be stopped by any obstacle,
even simple politeness and regard for a decision taken by a Congress.   He is too much accustomed to what is called in the poli-
tician language "lobbyists' methods," to have paid any attention
to what was unanimously adopted in London.   He saw at once
that the U. S. Geological Survey would derive renown and probably rewards; that it would please the Secretary of the Interior,
his direct Chief, to open solemnly the meeting in the capital of
the Union, and that it would make a good case before the Committee of Appropriations in the U. S. Congress, to obtain more
money, and at the same time silence his scientific opponents, if
any dared to attack him in his four Surveys [Geology, Topography, Irrigation, and Ethnology].

By a series of successful manœuvring Mr. Powell carried tri-
umphally his plan, and the International Congress of Geologists
convened at Washington in August, 1891, instead of Philadelphia.
Only the President and the other leaders of the London Congress
of 1888 thought the process of passing over their heads and put-
ting aside the resolution voted by the Congress rather high
handed, and a little out of the way of proper regard due to gentlemen and to science ; and in a body, all the officials of the International Congress of London, and all the officers of the Geological Survey of the United Kingdom of England and Ireland,
of Canada, and of India, kept out of the Washington meeting.
English geologists were represented only by a handful of members — ten and three Canadians — and not a single one among
them having a reputation and an acknowledged prominence in
geology.

The number of members present at the Washington Congress
was less than at any of the other meetings ; and it was anything but
a scientific success. Generally foreign geologists were astonished

to see in what hands the U. S. Geological Survey was placed. But Director Powell does not care anything about scientific results, he only wanted to make a political impression, and he got it, and now he is making the most not only of the meeting but also of the gift of the *Prix Cuvier*, lately given — December, 1891 — to the Survey, as a precious reward for the work done by the U. S. Geological Survey. An explanation is necessary to understand the true meaning of the choice of the French National Academy of Sciences. The only member of that Academy, present at the Washington meeting. M. Gaudry, was much struck by the great collections of American fossil vertebrate which he saw at Washington, Philadelphia, Cambridge and New Haven. Being a learned paleontologist for Mammalia himself, and knowing the great rivalry and jealousy existing between Messrs. Cope and Marsh, he turned the difficulty of a choice, in true Parisian way, by giving the price to the Geological Survey and not to an individual. As the actual Geological Survey has absorbed and is the successor of the Geological Survey of the Territories and of the Survey west of the 100th Meridian of which Mr. Cope was the paleontologist, the Cuvier prize is bestowed as much on the excellent work done by Cope, as on the publications of Marsh. M. Gaudry took the only way not to wound the pride of the two contestants, at the same time seeming to reward the whole Survey. But to take it as truly an approbation of the United States Geological work by the French National Academy. is going far from the mark. It is only a reward for the work done in vertebrate paleontology by Messrs. Cope and Marsh and their numerous assistants.

International Congress of Geologists are made especially to compare the geology of one country with the others ; to see the most important localities for fossils, for classifications, and to be acquainted with what has been done already on the general geology of the country where they met. For the European and foreign geologist in general it was a great treat, to expect to see the stratigraphy of another continent, and be able to compare classification, nomenclature, faunas, and Geological Maps. So much has been said of large sums of money expended annually by the United States Government, that it excited no small curiosity to see practically and *in situ* the results arrived at.

The managers of the Geological Survey, in order to meet the expectations, and at the same time to cover as much as possible their past and present errors, thought that something grand must be done to impress favorably the foreign visitors; and at the same time silence all opposition; and having that special purpose in view, they instituted in the Survey a new division, under the name of "Division of Geologic Correlation"[1] showing once more their tendency to use big sounding words, and to render the geological language as unintelligible as possible to the non-initiated.

Never before a Geological Survey has thought of such a division as one of "Geologic Correlation." Everybody who has any claim to be a geologist knows how to make synchronism, how to find equivalency or correlation as it is styled by some of the officers of the U. S. Geological Survey. The principles used everywhere by every one worthy the title of Geologist, have been set forth and are exposed in Elementary treaties or Manual of Geology for students; but it is almost comical to see a Geological Survey, institute a special Division for it.

The general purpose of that Division according to Director Powell "is threefold: (1) To exhibit in a summary way the present state of knowledge of North American geologic systems; (2) To formulate the principles of Geologic Correlation and Taxonomy; (3) To set forth from the American standpoint the possibility or impossibility of using in all countries the same set of names for stratigrapic divisions smaller than those systems." Accordingly Director Powell has assigned the work as follows:

Pleistocene: T. C. Chamberlin.
Neocene: William H. Dall.
Eocene: W. B. Clark.
Cretaceous [including the Laramie]: C. A. White.
Jura-Trias: I. C. Russell.
Carboniferous: ⎫
Devonian: ⎬ H. S. Williams.
Silurian: ⎫
Cambrian: ⎬ C. D. Walcott.
Algonkian: ⎫
Archean: ⎬ C. R. Van Hise.
Correlation by Vetebrate paleontology: O. C. Marsh.
Correlation by Paleobotany: Lester F. Ward.
Résumé of North American Stratigraphy: W. J. McGee.
Discussion of principles of correlation: G. K. Gilbert.

[1] Correlation does not exist in any of the Dictionaries of Geological terms; and is a poor substitute for synchronism and equivalency.

It was expected and even announced that all those reports were to be distributed at the Meeting of the International Congress of Washington, August, 1891 ; but one Report was issued in time ; *Bulletin* No. 80. "Devonian and Carboniferous," by H. S. Williams. It is simply an essay written by a person unacquainted practically with the Carboniferous and Dyassic periods.

Since the meeting of the Geological Congress, two other Reports have been distributed ; Bulletins Nos. 81 and 82, by Messrs. C. D. Walcot and C. A. White. Both Reports contain an unusually large contingent of errors, a no small amount of partisanship and rather curious one-sided historical sketches. The authors are even unable to make at least a show of justice in their lists of papers by dates and by authors, suppressing two thirds of the papers of their adversaries, but at the same time, being very careful to give the title of their own papers, however trifling they may be. On the whole those Reports Nos. 81 and 82 are only grand glorifications of the two authors' poor and even erroneous works.

So far the three Reports distributed are anything but creditable to the Geological Survey, and certainly do not add to the esteem entertained of the value of its publication. The institution of a "Division of Geologic Correlation" is another mistake added to many others, and shows how inadequate are the persons at the head of the Survey to deal with the main problem of the construction of the Geographical map of the United States.

The constant preoccupation of the Survey to expose, before the scientific world its "principles," is childish in the extreme. That sort of class instruction, which would be in place in a school or a college class, has become such a nightmare with the leaders of the Geological Survey, that they have the queer notion to bring the subject before the International Congress of Geologists at Washington, to the great amazement and at the same time amusement of the foreign Geologists. Mr. Gilbert "opened the discussion by presenting a general classification of methods of Correllation." Is it possible that the Chief of the "Geologic Branch" of the U. S. Geological Survey committed himself to such school teaching before an audience of able geologists, who had crossed the Atlantic with the hope of seeing *in situ* and working for themselves and discussing the geology of North America? What a blunder!

A few quotations from how the "Biotic method of correlation" is made use of by some of the "Heads of Divisions" of the Survey, will give an idea of their methods — methods well known by every practical geologist. Only all depends on the capacity of the observer. If the person knows his business and is really a good observer, he will give exact interpretation and good synchronism and equivalency; if, on the contrary, he does not know how to observe, how to identify fossil remains, what special faunas are, what general lithological characters are, then his interpretation is at fault, and all his work is unreliable and deceitful. It is not a question of principles, all well known since William Smith, Alexander Brongniart, George Cuvier, Agassiz, Alcide d'Orbigny, von Buch, Barrande, etc., but only how to apply and make use of them with intelligence and good sense. All depends on the status of knowledge of the observer.

We read in papers published by some officers of the U. S. Geological Survey, who were charged with writing some of the Correlation Reports, the following extraordinary and curious correlations, and how fossil remains are determined and made use of.

(1) The genus *Microdiscus* is primordial and most characteristic of the Taconic system in America and in Europe. It was created by Dr. Emmons, on a specimen found by him in the Taconic slates of Augusta county, Virginia; and the name has been used ever since by all paleozoic paleontologists. Mr. C. D. Walcott regards the original specimen of Dr. Emmons as a young *Trinucleus concentricus*, or a young *Ampyx*; and he refers the Taconic shales of Virginia to the Hudson River group (*Bulletin U. S. Geol. Surv.* No. 30, p. 152).

(2) Restatement by Mr. Walcott of the long exploded error, made many years previously, that the *Athops trilineatus* of the Taconic slates, is identical with the *Thriarthus Beckii*, and characterizes the Utica slates (*Albany Institut*, Vol. X, p. 23).

(3) Mr. Walcott, against a remarkable paper of Dr. Holm, referred to the genus *Elliptocephalus*, called by him *Olenellus*, a Newfoundland trilobite of another genus, called *Holmia* by Mr. Matthew; and according to his wrong determination, Mr. Walcott synchronised the "*Holmia* zone" of Newfoundland, with the "*Elliptocephalus* zone" of Georgia, Vermont (*Tenth Annual Report*, pp. 634-640).

(4) Mr. C. A. White has identified the *Gryphœa Tucumcarii*, a most characteristic fossil of the *Gryphœa dilatata* type of the

Jurassic period, with the *Gryphæa Pitcheri* the most characteristic fossil of the Neocomian or Lower Cretaceous, and has in
consequence of his wrong determination referred all the typical
Jurassic strata of America to the Cretaceous period.

(5) Mr. C. A. White has determined some fossil bivalves of
California as belonging to the genus *Aucella*, first found in the
Jurassic formation of Russia; and he had mixed together true
Aucella with at least one, perhaps two others, very different and
distinct genera (*Monotis* and *Avicula ?*); and as a consequence
of his principles of "Biotic method of correlation," has referred
the Trias period of California, not even to the Jura period but to
the Neocomian, making a double mistake (*Monograph XII,
U. S. Geol. Surv.*, pp. 196, 226).

Many other quotations might be made of how Messrs. Walcott
and C. A. White make use of the "Biotic methods of correla
tion." The U. S. Geological Survey is especially anxious to express its opinion that lithological methods of correlation cannot
be safely used, even in passing from province to province. The
general lithological characters for finding good and exact synchronism at great distance is entirely out of the reach of the
practical knowledge of the whole Corps of the U. S. Geological
Survey. It requires such an amount of practical research in the
fields of both hemispheres, that very few observers are trained
sufficiently to know how to use them. Then the Geological Survey being unable to handle properly the lithological characters,
not only ignore them, but declare them worthless. Another
instance of the conditions under which geology is made use of
by the Survey, and how they handle geological questions.

Some may be inclined to say: "Errors and mistakes of details
are inevitable in a Geological Survey." It is very true. But
there are gross errors and inexcusable mistakes which cannot be
made with impunity and produce extremely grave consequences;
and all those quoted in this pamphlet and many others not
quoted are so enormous and so important in their result, that the
whole stratigraphic table of American classification and nomenclature is turned topsy-turvy, with such gigantesque confusion,
that the list used by the United States Geological Survey is almost
comical.

CONCLUSION. — All seem to conspire toward a dictatorship of
the Director of the U. S. Geological Survey. After the discour-

aging leadership of Messrs. Hall and Dana, American geology has passed into the hands of the U. S. Geological Survey and its ambitious Director. It is a great misfortune. Geology wants freedom, scientific honesty, and a great deal of good practical work done in the field, in all typical localities, and in all difficult areas of both hemispheres.

The ordeal through which American geology has passed during the last forty-five years is demoralizing in the extreme and most detrimental to real progresses. The Geological Survey, instead of improving on the past, is on the contrary aggravating the position by a system of bounty and subsidy to professors of Colleges, State Surveys, which place at his mercy almost all the young geologists of the present day. A stop must be made before long, if American geology shall make a figure in the scientific world, and be placed in its proper position. Costly publications, with a great number of illustrations, and enormous and extravagant editions to distribute, do not constitute progressive and good work in geology. It is just to recall to memory that all the best and most original papers, which have founded all the branches of Geology, were published modestly, in very cheap form and at very little expense.

To conclude there is no more time to lose in such a costly and almost childish experiment. It is fully time to reform and reorganize the United States Geological Survey ; to stop all injudicious and extravagant expenditures ; and try to get competent geologists for the work. A new act should be enacted by Congress ; a Committee of Congress appointed, and advice asked, not from scientist-politicians, or hangers-on for a good salary, or scientists not well acquainted with the works required to construct a great Geological Map, but from true practical geologists, domestic as well as foreign, having a world-wide reputation.

The costly experiment has lasted long enough to prove the total inefficiency of the Director and of some of his principal associates. It is high time to separate from the Geological Survey the Topographical and Irrigation Surveys and the Ethnologic Bureau, and to see that the government gets in good work,— the value of the large sums it expends so generously. Of the three millions of dollars already paid for pure geological work, without counting the money expended for the Topographic, Irrigation and even Ethnologic works, more than two-thirds might have been saved and much better results obtained at the same time. All scientific flourish and demagogue utterances may be dispensed with with advantage to the true progress of science in America and to the United States Treasury. Able men are not wanting in the United States for such an organization as a Geological Survey. But proper men must be chosen with great discrimination and tact; for no one will think for a moment to ask a house and sign painter to make pictures as valuable and as perfect as those of Millet, Meissonier, Gérome, Ingres, Troyon, Rousseau, Rosa Bonheur, Courbet, Français, etc.; and the idea of expecting valuable, exact and truly good geological papers and geological maps from persons almost uneducated in geological and paleontological science is altogether preposterous and impracticable.

The ordeal through which we have just passed during a decade is too expensive and too improductive for a first trial. Let us now go really to work if we want a reliable "Geological Map of the United States."

www.ingramcontent.com/pod-product-compliance
Lightning Source LLC
Chambersburg PA
CBHW022037080426
42733CB00007B/865